LETTERS HOME

A Paratrooper's Story

by Harland "Bud" Curtis
U.S. Army Paratrooper 1943-45 As Recorded

by L. Vaughn Curtis, Ed.D., son

Copyright © 2010 L. Vaughn Curtis
Renewed © 2012 L Vaughn Curtis

Cover Image 2012 David Nibley
Editing: H.L. Curtis, T.L. Curtis

All rights reserved. No part of this publication may be reproduced, stored in a retrieval system, or transmitted in any form or by any means. This includes electronic, mechanical, photocopying, recording, or otherwise, without the prior permission of the author. Requests for permission to photocopy any portion of this book for any use should be directed to www.eckohousepublishing.com

Printed in the Unites States of America.
First printing: June 2005   3rd ed. June 2010

ISBN: 978-1-4276-5030-6
Library of Congress Control Number: 2010933703

Proudly Published by
ECKO House Publishing
Sandy, UT 84092

Contact us at www.eckohousepublishing.com

This book is dedicated to the men of the 517th Parachute Regimental Combat Team, (PRCT)

"Battling Buzzards"

They never wanted to go to war, but they did.
Their country called and they responded.
They put their country above their own lives.
They volunteered to be Paratroopers!
They became the most elite fighting men the world had ever known.

Airborne!

H. L. Curtis

1

All Paratroopers wore an American Flag on their left shoulder during the Invasion of Southern France, August 15, 1944.
This is the Flag Bud wore on his uniform.

Combat Streamers that flew from the 517th PRCT Regimental Flag.

CONTENTS

Foreword ix

Chapter 1
How All of This Started...1

Chapter 2
Initial Training at Camp Toccoa, Georgia with the 517th.....11

Chapter 3
Letters from Camp Toccoa...17

Chapter 4
Transfer to Regimental Headquarters, Camp Toccoa 35

Chapter 5
Transfer to Camp Mackall, North Carolina 43

Chapter 6
Finally, Jump School..49

Chapter 7
Return to Camp Mackall 59

Chapter 8
Tennessee Maneuvers...87

Chapter 9
One Final Trip Back to Camp Mackall 95

Chapter 10
Transferred Out of Regimental Headquarters....................103

Chapter 11
Point of Embarkation (P.O.E.) 111

Chapter 12
Overseas Combat Duty 117

Chapter 13
The Battle of the Bulge - Belgium and the Final Battle - Germany 185

Chapter 14
The War in Europe Ends 223

Chapter 15
Home and Readjustment to Civilian Life 239

Epilogue 243

Appendix A 259

References 261

Index 263

About The Author 265

FOREWORD

By
L. Vaughn Curtis, Ed.D.

In the late 1950's when I was 8 to 10 years old my grandmother, Olive Luceil Curtis would take me out to the sun porch of her home in Long Beach, California. There it was warm and comfortable, and an enjoyable place to sit and listen to my grandmother read to me. She would retrieve an old brown accordion folder, open the lid, thumb through the various files, then select a specific letter and take it out. These letters looked very old to me. She told me these letters were from my Dad that he had written to her during World War II, from 1943 until 1945. I didn't really understand how important these letters were to her (and would be to me in later years) but she enjoyed reading them to me and telling me about all of my dad's adventures in the Army. In later years I found out my dad had written over 130 letters to his mother. Today, I have every one of those letters and they are now my treasures.

My Dad was a very strict father, and at times exercised his authority with corporal punishment. I would become angry with how my father was treating me, and would flee to my grandmother for sympathy. She told me the reason for his anger and having a short temper was because of the things he suffered in the war. She explained that after my Dad had gotten home from the war the family was sitting at the table eating dinner when some loud noise would occur. My Dad was under the table before anyone realized what had happened. My Dad was subjected to many, many hours of German artillery shelling during the war. I didn't understand what my dad had to go through in the military. My dad would never talk about it, but through my grandmother, I learned he was an Army paratrooper. She would read to me about his experiences in paratrooper training, then his combat experiences. I began to understand a little of some of the horrors my Dad suffered in the Army. I began to wonder if I could ever be as tough as my Dad? I concluded that I would never be as tough as he was. As my grandmother would begin to read one of his many letters to me I would open an old bread yeast can that she had. Inside were different mementoes my Dad had gathered while in the Army. There was a piece of German shrapnel, a German Army uniform collar insignia, a German patch with the Swastika and eagle, the kind you see sewed above the right jacket pocket of a German uniform. There was a round German medal that was black in color and had a German inscription on the back of the medal. I never knew what it meant. Other things in the can were different U.S. Army insignias, and I enjoyed looking at these things over and over again as she read to me. Although she never read all of the letters to me, she would select specific ones that she thought I might be interested in hearing. Such as his first parachute jump and combat jump. I began to realize what a unique man my Dad is. I understood that not everyone could be a paratrooper, and only a select few were destined to become the elite of the U.S. Military.

In those early days of my life it seemed natural that my Dad would write letters home. It was the only way people could communicate with each other when away from home. Telegrams and long distance telephone calls were very expensive in those days, and were only used in times of emergency or special occasions such as Christmas, a birthday, and of course Mother's Day. There were no such things as cell phones, emails, text messaging or fax machines. There were only letters and postcards.

Airmail letters cost 6 cents and postcards were 3 cents. Most letters from my Dad who trained in Toccoa, Georgia, then Fayetteville, North Carolina took over a week to get home to Long Beach, California. It took the same time for my grandmother to respond back to him. Then when he was overseas in combat it would be weeks and sometime months before a letter was received. In his letters home my Dad would express his displeasure with the Army mail system, especially the V-mail system. This mailing system was free and the letter would be photocopied, then reduced in size and sent home postage free. These types of letters took longer to be received by families back in the states.

It is difficult for us in such a modern world to understand what was accepted as the only way to stay in touch with another human being. Letters! What a wonderful way to communicate. Hand written letters that were personal between two real people showing the distinctive manner of who these people really were for that specific period of time. Could they spell correctly, make complete sentences, and express their thoughts to others with clarity? My Dad was a very good letter writer for an 18 year old, as you will see. He wrote seventy-four (74) letters home while he was in training. Fifty letters during combat and another six letters from France after the war ended. He wrote 130 letters home that have lasted over sixty-two years. They are here today to be read and re-read again and again. I am so grateful to my grandmother for saving these letters.

As I grew up, and became older I forgot about the letters. I assumed my grandmother must have put them away because I never saw them again after about the age of ten. Then in 1989, my grandmother passed away. Their estate remained untouched until her husband, my grandfather passed away in 1992. Now their two sons, my Dad and his older brother Bert had the chore of liquidating their parent's estate. My Dad and his brother divided their parent's possessions and sold their home. My father never mentioned the letters after the estate was liquidated. From 1992 until 2004 the letters were not even thought of. Perhaps they were thrown away, and lost forever? No one seemed to know or for that matter cared.

Then in the year 2000 some World War II movies began to reappear that depicted the experiences of Army paratroopers. Specifically the movie "Saving Private Ryan" and the HBO series, "Band of Brothers." For Christmas 2002, my brother Tim sent me a copy of this series, "Band of Brothers." I began to understand what my father went through in the paratroopers. "Band of Brothers" is my Dad's story. Although he was not assigned to the 101st Airborne Division, it is a story about all World War II paratroopers.

In July of 2003, my father told me of a reunion in Toccoa, Georgia, that was going to be held to thank the World War II veteran paratroopers for their service to our country. I began thinking that perhaps this might be an excellent opportunity to take my Dad back to his old training ground, and hear him first hand tell the stories of his paratrooper training. I quickly telephoned my brother Tim and told him of my idea. He thought it was great and we made plans to take Dad to the "Paratrooper Reunion" in Toccoa, Georgia. We all flew to Atlanta, Georgia, and traveled by car to Toccoa, Georgia. There we saw the areas where my Dad trained and the famous "Currahee Mountain" as depicted in the movie "Band of Brothers." We were amazed our father ran up and down this mountain day after day.

My brother and I decided we wanted to run the mountain. We changed into our jogging clothes and ran only a portion of the mountain. We started at the top and began to run down the mountain. We found the dirt road very steep and rugged causing us to lean back to keep from falling forward. This caused our legs to hurt. As we jogged the road began to be level and we would run for a short time with ease. Then the steep mountain road began another long vertical ascent. Now we were running up a very steep rugged dirt road and then descending downward again. It was long and grueling running up and down the mountain. About three-fourths of the way we stopped got into the car Dad was driving, and rode down the rest of the way. As my brother and I sat there panting out of breath, we wondered how these men could run this mountain day after day?

While visiting these sites the people of Toccoa were just wonderful to us. They did everything possible to make us feel welcomed. While in Toccoa Dad asked if we could go to Fort Benning where he had learned to jump from airplanes. Fort Benning was only four hours away from Toccoa. We agreed to drive there. What a remarkable experience we had as we saw the jump towers where he learned the skill of parachuting. The soldiers at Fort Benning treated him like a hero. Our entire experience was one we would never forget.

As we returned home a new bond between Dad, my brother, and I developed. We continually talked about his paratrooper training. My brother and I would often quiz Dad about his experiences. Some things he remembered as if it were yesterday. Other things had gone from his memory. This made my father want to find his letters that he wrote to his mother during the war. He knew they bore the record of what he went through. As the years have passed many of the memories my Dad had of his military experience have faded. In a quest to find what had happened to him at specific times during the war, the letters to his mother had to be found, if they still existed. Over twelve years had past since his parent's estate was liquidated. Did the letters still exist? Only he would be able to answer that question.

In January 2004, my father went to his garage with the intentions of sifting and sorting through the many things stored there for the purpose of trying to find those letters. It had now been twelve long years since he settled his parent's estate. He wasn't even sure he had gotten the letters from his parent's home. Perhaps they didn't even exist. He could only search through the many items now stored in his garage. For three long days he spent searching for those letters. Finally, he came to a small travel case that was boxed shaped. It was 12" long, 9" wide and 5" tall with a handle. This case was more than likely a small travel case my grandmother used to carry makeup items when traveling in the 1930's and 40's.

In 1992, this travel case and many other items were just taken by my Dad, and put into his garage never knowing what was inside them, and not having the time to look through each item. Now, in 2004, he had time to investigate each item. My Dad took the travel case, opened the latch, and raised the lid. There inside were the letters he wrote home to his mother during the war from 1943 to 1945. They did exist, and the records of his war experiences were found. He quickly opened a letter and began to read, then another, and another. He spent time rereading most of the letters, and remembering those special times. It was as if he was back in time re-experiencing those days. After a few days he telephoned to tell me he had found the letters. I was elated and overcome with joy that after all these years the letters were still in existence. I was so happy that he had the determination to look for these letters. I had thought of these letters many times over the years, and especially when we visited Toccoa, Georgia, the letters came to my mind time and time again. I again remembered my grandmother reading them to me. I knew my grandmother would never throw these letters away. However, I was afraid that perhaps in the liquidation of their estate the letters might have been thrown away. Much to my surprise I was thrilled they were still here. My Dad quickly sent them to me by Federal Express and insured them against loss. For the few days in the mail we both waited with anxiety if they would arrive. On a cold winters day in February in 2004 I received the letters. I quickly opened the box and began to read them. I could hear my grandmother's words, and I felt as if I was again on the sun porch of her house hearing her read these to me. I was overcome with emotion. What a treasure to now again have. I was so grateful that these letters were still here. As I read them I said to myself, this information needs to be recorded. I thought of photocopying the letters. With 130 letters it would take forever, and be very costly. It came to me that I should transcribe them, and condense them into a book. A book about letters home: A paratrooper's story! Now I have done this, and have included historical facts that have been recorded from various books, especially the book "Paratrooper Odyssey." The 517th Parachute Regimental Combat Team, the "Battling Buzzards."

I offer to you the poignant, sincere, heartfelt letters of a loving son who wrote to his mother during the most horrific times of the twentieth century. This family of four was very close and not even a war could keep a mother and a son apart. These letters appear just as my father had written them with some standardizing of spelling and punctuation. I am amazed at the ability of a boy of eighteen who became a man very quickly to write with such deep conviction and description. These letters are a historical account of what actually took place. They are not a recollection of what my Dad remembered years later, rather an account of what exactly happened on that specific day of that year. Thank you Dad for these letters; thank you for the long suffering you endured in serving your country in most difficult times, serving in the toughest organization that was called on "To save the world." Your sacrifice and men like you will never be forgotten. As Tom Brokaw, a news commentator for NBC described, you were and are "The Greatest Generation" There is no one like you.

AIRBORNE ALL THE WAY!

Over 130 letters and cards found in a small travel case by Bud Curtis in February 2004. His mother saved every letter and card Bud wrote home from 1943 thru 1945.

Chapter 1
HOW ALL OF THIS STARTED

Harland "Bud" Curtis' High School Graduation Picture 1942

It was 9:00 pm, June 3, 2004, Bud Curtis and his oldest son, L. Vaughn, had just left the French Embassy in Washington, D.C. There the French government had just hosted a reception to honor one hundred (100) World War II veterans that had fought for France's freedom during World War II. It was a moving event. French officials told the American veterans they would never forget what America did for France, sixty years ago. The French officials thanked the veterans for all they had done to free France. Now after all this time Bud, First Battalion, and Walter Goforth, Regimental Headquarters of the 517th PRCT with ninety-eight other World War II veterans were going to be honored in France by receiving their highest military award. The "Legion of Honor." It was the equivalent of America's Medal of Honor. It was difficult to believe France would do this. What a great honor for two men of the 517th PRCT to be recognized by the French government along with other aging American war veterans. After the reception charter buses took all of the veterans and their escorts to the Dulles Airport in Virginia. To meet the veterans were Air France employees who gave special mementos to the veterans before boarding a charter flight direct to Paris, France. The flight left about 10:30 pm and took nine hours to arrive in Paris.

As Bud took his seat and fastened the seat belt his mind began to wander back to what had led up to this date, and how he was one of only one hundred World War II veterans to be selected to travel to France via Air France Airlines to receive the French Legion of Honor Medal in Paris France, on June 5, 2004. Bud was now within hours of receiving this great honor with his fellow veterans. In addition to this ceremony the French Government continued honoring these brave men of World War II by chartering a train that would travel the next day June 6, 2004, from Paris to Cairn. There the veterans were bused to the American Cemetery at Omaha Beach to listen to President George W. Bush then French President Jacques Chirac express how grateful they were for what American soldiers had done sixty years ago.

As we flew to France, Bud remembered when he was about 16 years of age and related the following, "I started thinking of all the crazy things I had done as a kid. I remembered one particular incident where from time to time my friends and I would go downtown to the Pike. The Pike was an amusement park built right on the oceanfront in Long Beach, California. There was a rollercoaster there (in those days it was called the cyclone racer) that was world renowned and known to be one of the most exciting thrills on the west coast, and perhaps the world. People were killed on this rollercoaster because the first dip was so steep. It was ninety (90) feet down at a fifty (50) degree angle, and the cars would attain speeds in excess of eighty (80) miles per hour. Many times the fatalities were drunken sailors from the near-by Navy base who would either stand up or just be thrown from the cars."

Bud and his friends (*cousin Bob Douglass and friend from school Roger Homer*) would purchase tickets to ride the rollercoaster and always get in the very first seat. Bud remembered, "The attendant would come by and push the safety bar down to hold us in the seat. When we left the station and began the long slow climb to the top of the track, my friends and I would release the safety bar latch and then hang our legs over the front of the car. As we reached the top it seemed like we hung there for many minutes, but in reality it was seconds, and all of a sudden, whoosh down they came going very fast, maybe a 100 miles per hour almost straight down. Our legs would vibrate back and forth and I would think that perhaps I would fall over the front of the car onto the tracks and be crushed. Once we passed the first dip my friends and I would get out of the front seat of the car and begin climbing back to the end car. People were screaming at us to sit down, but we kept moving back to the last car. By the time the ride ended we were at the very last seat. The attendant would look at us and hear what the other passengers said about us moving from car to car. The attendant would yell at us and kick us out telling us never to come back; but we did, and boy did we ever have fun. Why and how we never were thrown from the cars of the rollercoaster I will never know, but I enjoyed the excitement and the thrill of cheating death."

For years there had been the Naval Shipyard and Naval Base in Long Beach where ships were built. This provided good employment for the local community along with a major seaport for commercial and U.S. Navy ships that were docked in Long Beach. Many times while walking around the Pike with his friends or being downtown with his parents, Bud would see many sailors from the ships. These men were walking the streets with nothing to do, and usually looking for trouble.

Bud continued, "When my parents would take my brother and I downtown to shop, the streets from Ocean Boulevard to 6th Street were covered with sailors. I saw many sailors roaming the streets while they were on liberty (time off from work). Their obnoxious behavior, use of profanity, being intoxicated in front of my family while we tried to shop at the stores in downtown Long Beach offended me greatly, and I grew to hate the sailors because of their behavior.

Our world in Long Beach was quiet and beautiful. We lived close to the ocean and the weather was warm most of the time. In September 1939, I started my high school career as a brand new tenth grade student. I lived in an area where I was supposed to attend Wilson High School, located at Tenth and Ximeno Avenue, but I didn't want to go to Wilson High School. I wanted to go to Poly High School, located on Atlantic Boulevard and about Fifteenth Street because they had a Junior Reserve Officer Training Corps (JROTC) program there. Although I was unaware of what was going on in the world, our country was very patriotic and the military was an inviting prospect. So I requested to go to Poly and was granted permission. To get to Poly High I had to take a city bus from my home across town to Poly High School. Taking the bus every day required me to get up earlier than usual so I could be to school on time. This was just fine with me because I received a uniform and participated in military training. I loved being part of JROTC program. One day as I got off the bus and was walking home, three neighborhood girls saw me in my JROTC uniform. They followed me home and we talked. They told me I looked so handsome in my uniform. As we arrived at my house, I invited them to sit with me on our front porch and talk for a while. We sat on the large front porch steps of my parent's home. As we sat down, I had one girl on each side of me. The third girl had nowhere to sit and she asked, "Where should I sit?" Being smug and feeling I was something special, I said, "You can sit at my feet." This did not go over very well at all, and she became very upset. I learned a valuable lesson that day. Girls do not think what boys say is funny. Anyway we had a nice visit and then they went home.

As the tenth grade school year ended in June of 1940, I found it very difficult to continue attending Poly High School. Taking the bus across town to be able to wear an Army uniform and attend JROTC was just too much. When my eleventh grade (junior) year began in September of 1940, I made the decision to transfer back to Wilson High School where I could walk to school. The year started off okay, but I did not attend all of my classes and now I was short some credits for graduation. As school ended in June of 1941, I met a neighbor girl who lived across the street from me. She caught my eye and we began to date. Her name was Jill and she lived with her mother and brother.

The summer of 1941 was exciting, school was out and I had the entire summer to do what I wanted. I wasn't worried about high school credits and figured something would work out for me. Jill and I dated a short time and then one day she told me she was going to Utah to spend time with her father for the rest of the summer. I was not happy with this, and wondered what I would do for the rest of my summer vacation?

When school began again in September 1941, I continued to attend Wilson High School. It was my senior year and I was glad I would soon be through with school. Much to my surprise I was told I would not graduate. Over the course of my high school career I had missed some classes, and as I started my final year of high school, I was fifteen credits or three classes short of graduating. Although I knew I could not graduate on time I still attended school. As the first semester was just about over World War II began for America. It was Friday December 5, 1941, and school was over for the week so I had my weekend set for whatever I wanted to do. Saturday morning was spent helping my Dad with yard work at our house. Sunday was a day of worship, and my family belonged to the Church of Jesus Christ of Latter Day Saints (commonly called Mormons). We attended church every Sunday, but as a teenager I was bored with church. I was looking for more exciting things to do like driving cars, and dating girls rather than attending church. I respected my parents and always did what they wanted me to do. However, this Sunday morning December 7, 1941, would be different from any other Sunday I had ever known.

At 3:00 am California time the Japanese began their attack on Pearl Harbor in Hawaii. When we awoke that morning the radio was humming with the news of the attack. We later learned that the Japanese had attacked first with 183 bomber and torpedo planes then another 170 planes at 8:40 A.M. The Japanese had destroyed our American forces catching us completely off guard. They destroyed 328 aircraft, 8 battleships, wounded 1200 and killed 2400 people. My family and I, along with the rest of the world, were shocked. Later we heard on the radio President Roosevelt declare war with Japan and Germany. Earlier in my youth, in the 1930's, I saw President Roosevelt in downtown Long Beach driving by in a convertible with a long cigarette holder in his mouth. He made quite an impression on me. In six days, December 13, 1941, I turned 17, and I knew eventually this war would involve me, but just when and where I did not know. Once our Christmas break was over I returned to Wilson High School. A new wave of patriotism emerged and everyone began saving everything they could for the war effort. Food items and gasoline became rationed. I found the war was now affecting me personally, but I was still enjoying the latest, greatest, hits of 1941: Take the "A" Train, by Duke Ellington. You Made Me Love You, I Didn't Want To Do It, by Harry James. Boogie Woogie Bugle Boy, and I'll Be With You In Apple Blossom Time by the Andrew Sisters. Chattanooga Choo Choo, and Elmer's Tune, by Glenn Miller. Dolores, by Bing Crosby. Daddy, by Sammy Kaye. Oh Look at Me Now, Green Eyes, and Maria Elena, by Jimmy Dorsey.

In January things were not going well at Wilson High. I knew I could not graduate being three classes (five credits for each class) or fifteen credits short. I asked for, and was granted a transfer back to Poly High School. The reason this time for my transfer was that Poly High offered a radio science class that Wilson High did not have. I will never forget Mr. Farraund, who taught the radio class. This was the only class I really enjoyed. My radio science class was the reason I stayed in school. My older brother Bert was an electronic wizard. He built radio sets and other electronic devices. I wanted to be like my brother so I tried to follow in his footsteps by taking the radio science class at Poly High School. While at Poly High I met Garth Boyce. Garth and I became close friends. Garth once lived in San Francisco and told me how wonderful it was there. One day Garth came to me and said, "Bud, I have gotten us jobs selling shoes." The next thing I knew Garth and I were selling shoes every Friday night and Saturday. The money we earned was now being saved to leave Long Beach and go to San Francisco. In June 1942, my high school senior year ended. I did not graduate. I became bored with living at home in Long Beach, and I wanted to be on my own.

I was now seventeen and knew this could be my last carefree summer as I would eventually be drafted into the military. When school ended, Garth introduced me to a friend of his named Eddy Hunter. Eddy was a husky fellow with a large neck and strong arms. Garth and I both felt safe when Eddy was around so we invited him to go with us to San Francisco. In June 1942, the three of us left for San Francisco and rode up on a slow bus going the wartime speed limit of only 35 MPH. It took one day of travel to get to San Francisco. When we arrived we had no place to stay, so we slept in a bowling alley. The next day we went hunting for an apartment. There were very few to find. We began to worry we might not find one. However, by the end of the day, we found one at 925 Jones Street. The landlady never asked us our ages since all of us were wearing suits and ties. It must have made her think we were older. Now we had to find jobs. The three of us quickly started our job search. Eddy and Garth found jobs right away at different locations with Foster Fast Food Cafeterias. I still had not found a job, and continued looking until I finally was hired at the Stewart Hotel (known today as the Handerly Hotel) in downtown San Francisco. My new job was as an elevator operator. The hotel was across the street from the swank Saint Francis Hotel on Geary Street. In those days the elevators had to have an operator to make it go up and down. Trying to get the elevator to stop exactly at floor level was difficult. I had trouble at first getting the elevator to stop evenly at the floor. I would either stop an inch or two high or low. I usually said to the hotel guests, "Please step up or step down." My performance as an elevator boy must have been good because within two weeks I was promoted to "bellhop." This meant a lot more money. As a bellhop I would usually make $16 per day. Most adult men could only make $10 per day, and I was just 17. On a few good days I would make $100 per day. Garth and Eddy became very jealous concerning how much money I was making, as they continued to earn just a pittance in wages compared to what I earned. I told them why don't you guys try and get jobs as bell hops? They thought that was a great idea and so they did. Garth got a job at the Canterbury Hotel and Eddy got one at the Cliff Hotel on Geary Street. The three of us were now rolling in the dough. Our jobs were great and we earned lots of money from the tips we received from hotel guests. Sometimes collectively we had as much as $400+ per week. This was a great deal of money for three young men. We were living high on the hog and enjoying life. It seemed to me that leaving high school was the smartest thing I ever did. I was now a man (or so I thought) and on my own earning good money. Even though I had enough money, I still missed my family dearly, but I was not ready to go home yet. It was now September of 1942, and I decided that I wanted to get my high school diploma. I only needed three classes to graduate. I found out that Galileo High School, on Van Ness Street *(which was a mile or two across town)*, would accept me as a student. I left my high paying bell hop job at the Stewart Hotel to take a lower paying bell hop job at a smaller hotel at night so I could attend school in the day time. The schedule was rough. I was working all night and attending school all day along with completing the homework assignments. I had only been in school six or so weeks when my history teacher was absent. An obnoxious man was called in to substitute for him. I was sitting in class minding my own business when the substitute said, "Stand up and do not sit down until I tell you to." I asked why? He replied, "Because you were talking." I responded, "I was not talking." He shouted in a loud, mean voice, "Stand up as I directed." I told him he could go straight to hell! The other students sat there in shock and horror. They had never seen a student act this way before. The substitute shouted, "I am going to get the principal," which he did. In minutes the substitute and the principal appeared at the rear class door. The principal instructed me to come to the door, which I did. As I stepped into the hall the principal grabbed me by my right arm very tightly, and I told him, "Get your hand off of me." He squeezed tighter and I turned so I could get my left arm around his neck.

    I had him in a headlock and forced his face into a drinking fountain and turned on the water. Just then the bell rang and all of the students were now in the hallway observing this scene. I quickly released him and walked out of the school. I sat down on the curb and pondered what I had done. How did this happen. I knew I was tired from working all night and going to school all day. The substitute had no right to accuse me unjustly, but that did not change what had happened. I became concerned that the principal would somehow find out I was living in San Francisco with two other seventeen-year-old boys. I went to my locker, collected my books, and returned to the principal's office to turn in my books. When I walked into his office the principal had his head down saying, "I have never, I have never, experienced anything like this." I told him I was sorry and the best thing for me to do was to leave school. He agreed and I gave him my books and left (*Note: In 1946, after the war, I received my high school diploma, from Poly High School*). I went back to the Stewart Hotel, and asked for my day job back. They quickly obliged. It was soon thereafter that Eddy told Garth, and me he had been drafted. He told us he had to return home to Long Beach to be inducted. We were very surprised and our "Life of Riley" was coming to an end. Garth and I decided we would leave San Francisco and go to Florida where there was big bell hop money just waiting for us. We purchased tickets on the train for Florida and left San Francisco. The train stopped in Salt Lake City, Utah, where Garth had relatives and many high school friends. Once arriving in Salt Lake City, Garth wanted to get off of the train, and visit with friends for a day or two, then catch the next train to Florida. I agreed and we rented a $2.00 a night hotel room on the west side of Salt Lake City. Garth contacted his friends and the party began. I could not believe how many friends he had in Salt Lake. They never stopped coming over and the parties never ended. Garth was using up all of our funds. I had to get a job to keep us in the life style Garth had established for us. I was able to get on at the Kennecott Copper Mines cleaning the large ovens that were used for melting the copper. It was a hot dirty job and I hated it. I now earned $6.00 per day. This was quite a contrast from earning sometimes $600.00 a month in San Francisco. While I was in Salt Lake City I became concerned with the possibility of being drafted into the Army. I was almost 18 years old (17 years 11 months) and knew that my day of reckoning with the military was coming soon. I decided I would join the military. Growing up in Long Beach I had seen many sailors, and I thought I might join the Navy. I remembered the sailors from Long Beach were young men like me. Every Army soldier I had come in contact with were old men, or as I called them "old duffers." I didn't want to be in the Army with old men that reminded me of men my father's age. Everyone I knew was trying to stay out of the Army. It didn't seem to be a very smart idea to go into the Army. A guy could be on the front lines and get killed. I made up my mind to join the Navy where there were men my own age and there may be an opportunity to be safe and out of harms way.

To be able to join a military service after war was declared a person had to be 17 years of age, and had to have written permission from their parents to enlist. If a man was 18, he could not join, but had to wait for a draft notice. I was closing in on my 18th birthday fast, and so I quickly wrote my parents asking them to send me a written permission so I could join the Navy at age 17. I received written permission from my parents on my eighteenth birthday, Sunday, December 13, 1942. I quickly reported to the Navy recruiter's office the next day, Monday, December 14, 1942. I presented the letter from my parents giving me permission to join the Navy at age 17. I knew now I would be in a service that was much safer than being in the Army. I felt relieved that my worry about being drafted into the Army was over. Then the Navy recruiter told me he could not let me join because I was now 18 years of age. My birthday was just the day before. Couldn't they make an exception? It was only one day, but regulations were regulations. I missed the opportunity to be in the Navy by one day.

Working in the copper mines was a tough job. Garth and I grew tired of Salt Lake City, and wanted to go home to Long Beach, California. In December of 1942, we took the train to Los Angeles, and then got on the "Red Car." This was a large streetcar that went to Long Beach. Garth and I arrived back in Long Beach on New Years Eve. We were out of money, and began to walk the three miles to my home. We saw a man dragging a burlap bag. I asked him what was in the bag. The man responded bones for my dogs. I asked how far do you have to drag that bag? The man replied, "Another block." I said we have three miles to carry these suitcases before getting home. The man must have felt sorry for us and stuck his hand in his pocket and pulled out two dimes and gave them to us so we could ride on the seventh street streetcar. Once home my mother, father, and brother were very happy to see me, and I was very happy to see them. I was now eighteen with the draft board breathing down my back. I knew it would not be long before I would be drafted. For now it was just good to be back in my own bed, between clean sheets that my mother washed. To eat home cooking and to have ice cold milk every day from Mom's refrigerator was heaven on earth. As 1942 came to an end I loved to listen to big band music like: Don't Sit Under The Apple Tree With Anyone Else But Me, I've Got a Gal in Kalamazoo, Moon Light Cocktail, and A String of Pearls, by Glenn Miller. Somebody Else Is Taking My Place, and Jersey Bounce, by Benny Goodman. Jingle Jangle Jingle, by Kay Kyser. There'll Be Bluebirds Over the White Cliffs of Dover, by Kate Smith, Flying Home, by Lionel Hampton, For Me and My Gal, by Judy Garland. These were the hits of the day."

In January 1943, Bud went to work for the Pacific Borax Company in Willington, California. Bud recounted, "My job was to stand under a hopper and fill 100 pound bags with crushed borax. I would pick up the l00 pound bags and placed them on a large dolly until there were six bags. Then I would push them to a boxcar train. I and another worker would stack the bags in the boxcar. This was hard work and I was getting very tired of it. I wished I were back in San Francisco earning the big money as a bellhop."

One day in February 1943, the radio announced that the Navy would accept new recruits for a limited period of time that day. A friend and I quickly drove up to the Los Angeles induction center. There I went through the in-processing procedures to be in the Navy. I took the induction physical, and was told to sit in a specific location and wait to be sworn in. I was convinced that I was going into the Navy. This was going to be my best option to survive the war. The Navy was a clean and safe service to be in. My thoughts reflected fforts to join the Navy in Salt Lake City, Utah. Every time I thought about being drafted into the Army, all I could see in my mind's eye was old men. I didn't want that. I was ready to be a sailor. As I sat there waiting the door opened and a Naval Officer announced, "There will be no more men accepted into the Navy today." I was disappointed and returned home.

The next Sunday I attended church with my parents in the newly completed Park View Ward building on Ximeno Street in Long Beach, California. While there I saw an old friend who a few months before was my Sunday school teacher. He was dressed in an Army uniform. It was Willard Hill. He was now a Second Lieutenant and was wearing his Army officer's uniform. He was a paratrooper! He had jump wings on his uniform, and more impressive than that were his highly polished jump boots with his pant legs tucked into those boots. You could see your face in those highly shined boots. I had never been so impressed by a military soldier in my entire life. I was very impressed and wanted to know everything about the paratroopers. Willard told me how great it was to be a paratrooper. I believed every word Willard told me. I wanted to be a paratrooper! This type of adventure fit right in with my life style. I enjoyed danger. I challenged death on the rollercoaster (called the cyclone racer in those days) as a kid. I rode my bicycle sitting backwards on the handlebars using my hands to steer without seeing very well where I was going. I had moved out of my home at seventeen and I lived on my own. I did not take any guff from any person as I displayed with my school principal in San Francisco. I considered myself a perfect candidate for the paratroopers.

Finally, the day arrived and I received my draft notice on, March 24, 1943. The letter was from the Local Draft Board to report for induction. I was order to report to the Pacific Electric Station, West Ocean Boulevard, Long Beach, California at 6:45 A.M., April 7, 1943. I reported for my induction physical, and once again thought about joining the U.S. Navy. I knew being in the Navy would be a good way to survive the war. I remembered Willard Hill and how exciting it seemed to me to be in the paratroopers. Joining the paratroopers would mean the roughest, toughest training there was. I would be on the front lines in direct combat with the enemy.

I started the testing and physical examinations at the induction center in Los Angeles. I was standing in line waiting to report to a solider at a desk. As I approached the soldier in charge at the front desk, he asked me which service I wanted to join? I hesitated. I didn't know what to say. I had pretty much made up my mind that day to join the Navy, but after speaking with Willard Hill at church I knew the paratroopers would be exciting and a real challenge. Paratrooper training would be dangerous and require a great deal more discipline than the Navy. Why would I want to take such a risk? The Navy would definitely be safer, and perhaps I could even be stationed right there in Long Beach, California. The soldier once again asked me, "Which service do you want to be in"? I quickly replied, "The Navy." The solider told me to go sit over there with the other Navy recruits. "You are now in the Navy", he said. I was finally in the Navy, which was no easy task. Everyone wanted into the Navy. The Navy had a quota. They only took a few men each day. Not like the Army or Marines who were taking every man they could get everyday. Young men being inducted knew the Navy was fairly safe and there would be no direct combat with an enemy solider. It was the smartest thing I ever did, or so I thought.

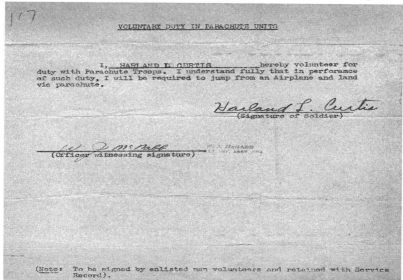

For five whole minutes I sat there with the other Navy recruits. I began to doubt my decision. I kept saying to myself, "What have I done, I can't go into the Navy, I want to be a paratrooper. I want to be like Willard Hill, What am I going to do? I don't want to be a sailor. I hate sailors." I mustered the courage and got up from my seat and went back over to the soldier and said, "Sir I want to change my branch of service." The Marine recruiter interrupted him and said, "Of course he wants to change services. He wants to be a Marine." I quickly said, "No, I want to join the Army. I want to be a paratrooper!" "The recruiters yelled at me, "Are you nuts?" Okay, if that is what you want to do, go over there and sit with the Army fellows." After in-processing that day I was sent home and told to report for duty in a week. I came back and report on April 14, 1943. I was transported a few miles over to Fort MacArthur that was just across the Long Beach Harbor in San Pedro. There I stayed for about a week being issued Army uniforms. It was there that I volunteered for the paratroops. My request was process promptly, and I soon found myself boarding a train bound for Toccoa, Georgia. My journey to become a paratrooper had begun. In 1943, I remembered the hit parade of big bands, my favorite hits were: Juke Box Saturday Night, That Old Black Magic, by Glenn Miller. You'd Be So Nice to Come Home To, by Diana Shore. Boogie Woogie, In the Blue of Evening, by Tommy Dorsey. Why Don't You Do Right, Taking a Chance on Love, by Benny Goodman. Comin'in on a Wing and a Prayer, by The Song Spinners.

As soon as Bud arrived at Fort MacArthur, California he quickly signed this form volunteering for parachute duty. He was soon on a train bound for Camp Toccoa, Georgia where his training to become a paratrooper began in April 1943.

The most thrilling attraction for over 4 decades, and billed as the greatest ride on the earth. The "Cyclone Racer" was a dual track wooden roller coaster built in 1930. This was the only dual track roller coaster in the United States. The dual cars would leave the station side by side and then plunge 90 feet down at a 50 degree angle attaining speeds in excess of 80 miles per hour. Many drunken persons or saliors lost their lives as they were flung from the cars. Bud, friends, sons, and over 30 million people rode the Cyclone Racer before it was demolished in 1968.

CHAPTER 2

INITIAL TRAINING AT CAMP TOCCOA, GEORGIA WITH THE 517TH

"Battling Buzzards"
517th Parachute Regimental Combat Team (PRCT) Unit Patch

Howard Hensleigh, (2009) tells us who designed the 517th Unit Patch and how they got the name "Battling Buzzards." Howard said, "The term "Battling Buzzards" has been with us for a long time. It was coined by Dick Spencer, an Iowa University classmate and 3rd Battalion warrior. He published the words "Battling Buzzards" in a booklet after the Italian campaign. Dick also was the cartoonist that developed the patches of the 517th PIR, the 460th Parachute Field Artillery Battalion, and the 596th Parachute Engineers Company."

In 1942 & 43, Bud was enjoying a carefree life with nothing more to worry about than his youthful travels all over the western United States with his friends. During this same time period Louis A. Walsh, Jr. was pioneering paratrooper warfare. He was one of the first men in his graduating West Point Class of 1934 to be promoted to full Colonel and received his Eagles in 1942. To be promoted to full colonel in eight years was unheard of, but these were different times. These were war years. Bright young men with visions of how to conduct new styles and types of warfare were desperately needed. Louis Walsh was one of these kinds of men. He was one of the very first men to become Airborne qualified, and to pioneer the concept of combat paratroopers.

The Army recognized Colonel Walsh as one who would be a leader of paratroopers and sent him in 1942 to the South Pacific to observe the American soldiers and Marines fighting the Japanese. After 10 weeks he returned to the United States, wrote an eighteen page after action report that later became standard operating procedure (SOP) manual for the newly organized 517th Parachute Infantry Regiment.

Astor (1993), author of Battling Buzzards remarked, "Colonel Walsh was sent to the south pacific to learn from a Marine officer that there was a need to weed out the incompetents or emotionally unfit." As soldiers began to report to Camp Toccoa, Georgia, Colonel Walsh determined that all troopers would have "A will to win." He intensified the training to find these kinds of men.

The 517th PRCT began it's 33 month life in March 1943, and was eventually known as the "Battling Buzzards" as their unit patch displayed an irate vulture against a parachute with the numbers 517 written above it. The 517th was assigned to the 17th Airborne Division in the States, but would go overseas as an individual regimental combat team, that could be used anywhere they were needed to bolster the front lines. These men had to be tough and ready for the most intense combat the war could offer. The call went out across the nation for men to volunteer for the paratroops. For several months the new volunteers arrived at Camp Toccoa, Georgia, for temporary duty and screening into the paratroopers. Men at first began arriving at about two hundred per day. Then they came anywhere from fifty to one hundred and fifty men per day. With only nine officers and a handful of non-commissioned officers (NCO's) the new cadre of the 517th began to receive the new Army recruits from across the nation. Trucks were borrowed and men were picked up at the train station and taken to camp. Soldiers would be issued clothing and bedding, and with some help from the Post garrison. However, Astor (1993) found that "The officers not only escorted these initial recruits to the camp, but issued bedding and cooked their first meals served in the mess hall." As the men arrived they began to fill the First Battalion of the 517th. This battalion like others had a headquarters company, and three line companies, "A", "B", and "C." Second Battalion would have line companies of "D", "E", and "F." A few months later Third Battalion was organized with line companies of "G", "H", and "I."

The Battalion Commanders of the 517th.

Bud started in Second Battalion, "F" Company, and was interviewed by then Major Seitz. Bud transferred to Regimental Headquarters, communications section, and then transferred just prior to going overseas to Headquarters, First Battalion. LTC Boyle and Bud came close to being killed by friendly artillery fire from the 75th Infantry Division in December 1944 in Belgium. Both men remembered that day on May 27, 2004, in "Mailcall" # 700 (see page 199). In speaking with Colonel Boyle at the 517th reunion in Washington, D.C., on June 30, 2007, he stated, "I completed jump school all in one day at Camp Toccoa in 1942. I made all 5 jumps in one day and was "Airborne" qualified."

Bud Curtis arrived at the Toccoa train station about a month after the 517th was organized. It was around the 15th of April 1943. Bud was assigned to the Second Battalion, "F" Company, and would be one of the first men to receive this new style training devised by Colonel Walsh. Bud remembered very clearly the day he arrived. "It was raining and cold and between 3:00 and 4:00 AM." Army trucks picked up the new trainees and took them to camp. As soon as Bud arrived at Camp Toccoa, he was taken to his barracks where he was assigned a bunk. He was then marched off to "Chow" (breakfast) around 6:00 AM. Then back to the barracks. There the first officer of the 517th he met was Lieutenant John Lissner. He was a rugged man and had a commanding voice. He yelled at the men, saying, "If you think this is bad now it is going to get worse, if you think the Chow was bad, it will get worse. f you think you can make it through this you won't." And he was right. Many of the men dropped out. Bud also remembered Lieutenant John Alicki. He was also a rugged man and took pleasure in greeting the new troops with,
"All right, ya volunteered for parachute duty, now is your chance to prove ya meant it!" One night after the men had gone to bed a man had to go to the Latrine (restroom). Lt Alicki woke up everyone and made all of the men go to the Latrine. For Bud, Lieutenant's Lissner and Alicki's authoritative voices and commands were nerve shattering, but not enough to scare Bud off. He would rise to the challenge. Next he was taken to the medical facility for a physical exam of which he passed. Bud wrote home that only 3 of 20 men were accepted into the paratroopers that day. Then he was taken to the mock-up airplanes, where an old outdated parachute harness was placed on his back. Lieutenant John Lissner, Camp Toccoa, GA, 1943

He climbed the ladder to the mock-up airplane. There he and other candidates were told to hook up their static lines to the steel cable, stand in the door and JUMP! This was the first day. Wasn't there any training required before he had to jump? Why was he forced to do this? It was frightening. How anyone could be required to jump out of a mock up airplane with no training. It seemed crazy. Many of the new paratrooper candidates felt just that way and hesitated at the door and could not jump. They were moved off to the side and disqualified from airborne training, and mustered out to the "straight leg" (ground) infantry. Perhaps Bud saw what was happening to these men, or perhaps his daring life style as a youth gave him the courage to jump right away. He didn't hesitate, but jumped right out and slid down the cable to the ground. Bud remarked in later years that, "The men who were mustered out where put on K.P. (kitchen police) for weeks before they could be processed out to their new units." That one fact kept Bud going he did not want to be put in a disgraced status of being a K.P.

Colonel Walsh ordered this kind of first day training to determine which soldiers were prepared to fight without waiting to see what other soldiers would do. He needed men who were quick thinkers and could make decisions quickly. Bud Curtis possessed those qualities. Colonel Walsh required men who were physically fit with strength in infantry tactics. Every morning immediately after reveille, the entire outfit from Colonel to Private fell out into formation for a two mile run around the camp. In the afternoon after lectures and field training the men fell out again, but this time the run was up Mt Currahee. A three and one half mile run from hut to mountain top. The road up the mountain was steep and difficult. Loose gravel and rocks lined the trail making it hard to keep ones footing. Running down the hill was equally difficult. It was so steep the men had to lean back in their running which caused extreme pain in their legs. Anyone falling out of these runs was quickly disqualified from airborne training. During one of these runs, Bud Curtis remembered a trooper from the 517th who collapsed and fell out of the run. Bud remarked, "This trooper would never have to worry about being disqualified. He had dropped dead from exhaustion." The pace increased for the others. There was no backing off or changing the training.

Little did the men of the 517th know what was in store for them when they finally got into combat. Gerald Astor (1993) reported in his book "Battling Buzzard", The 517th PRCT served more days on the front lines than any other combat unit in World War II. Ninety-four consecutive days without relief. No other unit would be able to claim such a distinction.

The arrival at Camp Toccoa was a shock for all men. They were now candidates for acceptance into the 517th Parachute Infantry Regiment (PIR). From now on everything would be done together, at the double time. If one man had to use the latrine all men would fall out into formation and use the latrine together. It was teamwork. There training was the toughest in the world. Each paratrooper would be equal to at least SEVEN regular soldiers. On his second day at Camp Toccoa Private Curtis was told to report Major Richard Seitz's office in one of the huts. He ran to the Major's office, and knocked hard on the door frame with his knuckles. "Sir, Private Curtis reporting as ordered." Major Seitz invited him in. Bud stood at the rigid position of attention while the Major interrogated him, trying to intimidate him and to see if he would break. "Why do you want to be in the paratroopers?" asked the Major. Bud responded, "I want to kill Germans, Sir." "You want to see their blood and guts Curtis?" Came the reply. "Yes sir!" Said Bud. More questions were asked then the Major said, "Is there anything else you want to tell me Curtis." For a split second Bud's mine began to think back to all the things he had done, and wondered if the Major knew about his past life. He quickly said, "No Sir." Bud was ordered out of the office. He had passed his interview. In speaking with now Lieutenant General Richard Seitz, USA,

Retired, at the 517th PRCT reunion in Washington D.C., on July 1, 2007, General Seitz said, "I interviewed 3,100 men and only allowed 700 into the 517th. Headquarters was putting a lot of pressure on me to get the battalion staffed, but I was looking for men with spirit and heart, not muscles." Bud was well on his way to becoming one of the most elite fighting men in the world. He was about to become a combat paratrooper with the 517th Parachute Regimental Combat Team (PRCT).

Colonel Walsh had given strict orders to screen the men very closely, and if the officers saw any flaws the soldier was washed out immediately. Higher headquarters took exception to these tactics and felt the 517th was rejecting too many men and the selection process was taking too long, but Colonel Walsh's standards would not be changed. Once a man got into the 517th he did not leave. He had gone through too much to be washed out of this outfit.

Now this is Bud's story about his experiences in the 517th, as recorded in letters to his mother. Bud came from a very close family and wanted to stay close to them even though he was thousands of miles away. In 1943 phone service was limited. There were no cell phones, text messaging, or email available. There was only the U.S. Mail. Letters took five days to California and five days back to Georgia. It was difficult to know what was going on at home except to receive these precious letters. It was the only way to communicate with family. This is his story through the letters he wrote home from the start of his training at Camp Toccoa, Georgia, to combat in Italy, France, Belgium, and Germany. A soldier's story of how it really was in the toughest combat unit America had to offer.

Mt Currahee, Toccoa, Georgia, picture taken September 25, 2003

Lt John "Boom Boom" Alicki, one of the first officers that Bud met when he arrived at Camp Toccoa, GA, in 1943. Later Lt Alicki became the officer in charge of the Demolitions Platoon.

Chapter 3

LETTERS FROM CAMP TOCCOA

April Our first day in the paratroops will always be remembered for those leaps from the high tower-mock-up.

Bud was on his way by train from Fort MacArthur, California. He had just volunteered for one of the most elite units in the America Army, but why? Willard Hill was a big part of it. Bud thought his uniform looked very impressive. Ruggero (2003) sheds some light on why men like Bud did what they did. "The Airborne reputation was one of the hardest assignments in the Army. Recruiters spread a clear message that only the toughest men would make it through the training." To win the coveted silver paratrooper wings a man would have to be fearless. Well, Bud thought he had the "right stuff" knowing full well that this was a highly risky business. Bud and the other new candidates knew the mission of the airborne was to jump from an aircraft in flight behind enemy German lines, at night. Each man hoped his parachute would open, hoped he did not break a leg or back upon landing on the ground. Then if he landed properly and his luck didn't run out, maybe he would not be dragged across the ground by a billowing chute. Oh yes, all the while being shot at by German soldiers while trying to get out of his parachute harness, then get his rifle out of the carrying case. These men developed an "Esprit" that would be second to none.

Bud's first letter to his family was on April 27, 1943. He explained in detail his first day and what it was like. His tells his family only 3 of 20 made the cut into the paratroopers on the day he arrived. Bud explains about jumping from the thirty-five foot "Mock tower." Most individuals would look upon this experience as dangerous and terrifying. What did Bud say about it to his family? "Boy is it fun!"

Letter to the Folks from Harland "Bud" Curtis, Camp Toccoa, Georgia

April 27, 1943
Company F

Dear Folks,

It is Tuesday and have I been busy since I called you up in New Orleans. Right after I talked to you we got a train and were on our way to Georgia. We went through Mississippi right where the river empties into the Gulf of Mexico. Then we went through Alabama (the train would stop periodically and in the evening Bud would see lighting bugs. He was intrigued by these little bugs that he had never seen before in his life) and finally early Monday morning we got to Toccoa (we had to travel on a passenger train because we missed the troop train in New Orleans by about 15 minutes so I didn't have any sleep from Friday until last night). Now here is the joke. As soon as we got to camp here we ate and then started the day by taking a physical exam. Well believe it or not I passed everything, I don't see how, but I did. If you could have seen how many flunked out you wouldn't believe it. The average was about 3 out of 20 that made it. Oh the two fellows I've been palling around with since Fort McArthur, made it too.

Well then we went out to a thing called a mock tower. It is about 35 feet in the air and is built like a middle section of an airplane, with a door on one side. Well you put on a parachute harness (without the parachute) and then you go up the ladder to the door where an instructor hooks you onto a pulley attached to a long cable. Then you get in the position at the door. There is a certain way you have to jump out; you don't just dive; there is a real technique to it. Well anyway you jump and count 1000, 2000, 3,000 as loud as you can and as you holler 3,000 you are at the end of your line and with a big jerk you are stopped. Then the pulley slides about 50 yards in a down ward sloop to where you can touch the ground. Boy is it fun!

Well that is about all we did yesterday. Oh and you don't do much walking in the paratroopers. <u>YOU RUN</u>! Boy, we're up in the mountains (is it pretty) big trees, and clear skies. It is just like camping out.

Well today the first thing we did was to jump from the mock tower. Then we had a personal interview by a Major (Seitz). Boy is he a swell guy. About 30 years old. Well I walked in. Saluted him, told him my name. He asked me how I liked the paratroops. I told him "PERFECT." He said he was organizing a bunch of paratroopers for actual combat overseas in about 9 months from now. He asked if I wanted to be in it. If you could have heard him, the way he said, "If you do come into my battalion, you are going to work, and work hard, and when I'm through with you, you'll be a man, and the very best." (Major Seitz also told Bud we are going to go to Japan and kill the Japanese, is that what you want to do? Bud responded, "Yes sir!") Then if you could have heard him say. "Write your folks and tell them you are in my outfit, and that you'll be overseas in 9 months, and tell them that I'm personally going to bring you and all the others back <u>alive</u>." Well cold chills went up my back, and I would have done about anything he wanted me too. Well anyway I'm on the road to being a paratrooper. But I am going to have to work for it. I think I can get in the signal corps of this Major's outfit I am in.

Now, I'll let you know that these paratroops are no bed of roses. The food you eat here is nourishing, but the way it is thrown together, you would think that camp food I ate was a banquet. Well, I eat it anyway and I'm gaining weight. Well, for 13 weeks now we will be having basic training somewhere; it might be in North Carolina or is it South Carolina. Were just 16 miles from the border anyway (he later learned it was North Carolina). And so after we finish this 13 weeks I think then we go to Fort Benning where Willard Hill is at.

If I do come out okay through all this training you can go to your bridge parties and when all the mothers start bragging, you can top them all, because it is the toughest service there is too get into.

As soon as I get 5 jumps in I'm a full fledged paratrooper, and I get a furlough, and I'll see you then, but that will be a few months yet, so in the meantime I'll write as often as I can, but don't worry if I don't. It will be because I won't have time. They can keep you busy I found out that. Well the address is on the outside of the letter. Write soon.

Love Bud

Mock Airplane fuselages Bud and the other men jumped from to prove their courage to become paratroopers at Camp Toccoa, Georgia, 1943.

Bud's training was intense and fast paced. The Army was overwhelmed with all of these new recruits reporting for duty, and found it's self short of needed clothing. Bud refers to these shortages in his letters to his mother asking for personal items such as socks, towels, undershirts and hygiene necessities. Oh yes, he wanted to know if his Dad could send a radio. This would surely be a luxury that most could not afford to have.

Letter to Mom from Harland "Bud" Curtis, Camp Toccoa, Georgia

May 10, 1943
Company F

Dear Mom,

I'm just going to write a quick letter as I haven't much to say. What I want is if you can send me a large bath towel, some shorts, and those undershirts. I can get T-shirts here, but I can't get this other stuff. Ask Dad if he can get a hold of one of those penlights, and please send me some socks too. I could sure use the radio alright, but this is quite a ways off to send it. If you do send it, put it in a good box so that when I have to I can send it back as I'll be shipping out somewhere else in a couple of months or so. Gosh it is been raining cats and dogs all day; this is sure funny weather, it will rain, like a cloud burst for a while and then the sun will shine so hot you think you are in the desert, then it will shine and rain at the same time.

Try to find out Eddie Hunter's address for me will you? His folks live at the Rose Courts, Broadway and Atlantic. They live in apartment number H, so look it up in the phone book. And find out Gary's address too please.

Did my Company Commander send you a letter about me being in this outfit? I heard he sends everyone's folks one (what did he have to say?) I'll send home some money to pay for those things, I get paid on the 15th and I'll send some money for the postage if you will have them send it here. Well that is about all so give my love to everyone. Write often.
Bud

Bud's Company Commander, then First Lieutenant John E. McKinley, from "F" Company did send a letter to his parents on May 10, 1943, telling Bud's parents about the training and how Bud would become one of the world's finest fighting men. Note: This was quite a letter to send to the parents of an eighteen year old boy who was being turned into a man very quickly to fight for America's freedom. Colonel Walsh felt it important to have the families of these new paratroopers involved in their lives. If family members would encourage their sons to do their duty they would be better fighting men. Colonel Walsh ordered each company commander to write a letter home to each paratrooper's parents. Bud started his career with F Company and here is the letter sent to his parents by his company commander.

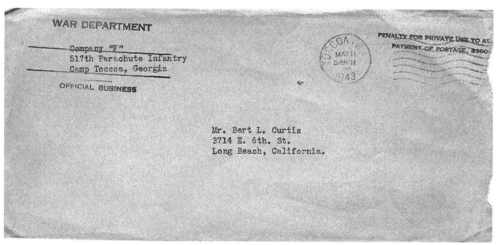

Envelope containing Letter sent to Bud's parents from Company Commander John McKinley

Company "F"
517th Parachute Infantry
Camp Toccoa, Georgia
May 10, 1943

Dear Mr. Curtis:

Your son has been assigned to this Company, and as his Company Commander I am writing to clarify any ideas you might have regarding the Parachute Troops and the men with whom he will be associated in his service to our country.

All Men accepted must be physically, mentally and morally in the highest bracket of American manhood, and they must have the capacity to absorb mental and physical training which will develop them into the finest soldiers in the world. You may well feel a high sense of pride in the fact that your son has willingly volunteered to serve his country as a "Paratrooper", a soldier whose distinctive uniform, bearing and soldierly appearance set him apart from other soldiers wherever he may go and gain for him the admiration and respect which the men in no other branch of the service enjoy.

The moral, mental and physical well being of your son and the other in this unit is carefully provided for. The men are required to keep the barracks in which they live spotless at all times; sanitation is impressed upon all personnel and barracks, meals, etc., are inspected daily by an officer. Free Medical and Dental service are available and all men are required to take advantage of them. Attendance at religious services of all denominations is encouraged and there is a chapel in the area with a Chaplain on duty at all times to care for spiritual needs. For relaxtion a Recreation Room is provided in the area and there are supervised athletics. Later on men will be permitted to go into town on week-ends where there is a fine U.S.O. building for their use. Your son will be put through a gradual physical training process which will develop him into a superb soldier at the peak of physical efficiency.

Your son has received the golden opportunity of being associated with officers that are highly specialized in this type of work. These men were hand-picked by the Regimental Commander who is a shrewd observer of leadership. Within our organization the enlisted man and Officer have a common interest, therefore they are closely united. It is this combination plus being the best trained, best equipped soldier in the world that will enable him to become the worlds finest fighting man.

Sincerely,

JOHN E. MCKINLEY
1st. Lt. 517th Prcht. Inf
Commanding.

Bud wrote his Dad on May 8, 1943 about Bob Hope coming to visit the camp and what a great show he put on. Bud and other soldiers probably never knew why Bob Hope came to entertain them. In the book Paratrooper Odyssey, the question was answered and recorded this way: "The Regimental S-2 Captain Albin Dearing was a sophisticated and cosmopolitan man with literary and theatrical connections. On learning one day that the well-known comedian Bob Hope was scheduled for an appearance in Atlanta, Dearing asked Colonel Walsh for permission to try to get him to make a side trip to Camp Toccoa. To the great surprise of everyone except Dearing, Hope agreed, and a few days later a C47 carrying him and his troupe put down after dark at a tiny nearby airstrip. Bob Hope was a quick study. His success is based in large part upon his ability to adapt to the idiosyncrasies of any particular audience. Enroute to the camp Colonel Walsh gave him a quick rundown on paratroopers and parachute training.

The entire regiment was assembled in the Post theatre. After being introduced with a not very original pun ("... I now present our last Hope...") Mr. Hope delighted his audience by pretending to struggle through an exaggerated pushup. The performance was brilliant. The troopers felt that if a celebrity like Bob Hope could take time to visit tiny Camp Toccoa, perhaps their efforts were being appreciated. In his own way Bob Hope contributed greatly to the war effort."

The First Edition of the Thunderbolt, Camp Toccoa, Georgia, August 1943, Volume 1, Number 1 had reporters writing about Bob Hope's visit to the camp.

"Bob Hope Livens Routine Grind With Hot Performance"

Colonna, Langford, Romano, Vague Help Keep Rookies Rolling In The Aisles" -

"The rookie left his barracks and marched down to the theatre. This night was to be his. Something different was coming into his sheltered Paratrooper's life. Why? Bob Hope was coming tonight. The Rookie would get his mind off the obstacle course. That run up Mt. Currahee could be forgotten — at least until tomorrow. What if his rifle wasn't clean. The sarge wouldn't see it — he was at the show.

Bob Hope reads inscription on leather cigarette case given him by Colonel Walsh as S/Sgt Oliver looks on

Lt Schmitz leads Songs then…

So what happens? He walks into the theatre and after nearly an hour's wait during which Lt Schmitz led the boys in a little song fest, the one and only Bob Hope walks out on the stage. They are going to put the new roof on the movie house any day now, but that is not the important thing. Do you know what that Hope did the very first thing? That's right — he got down and did push-ups. Of course his nice plaid coat got a little dusty and he collapsed after the fourth one, but that's not the point. The memories of the day's work were brought back to mind.

However, it is doubtful that anyone thought of anything but the fugitive from a tooth paste ad for the next hour and a half as he gave the boys all he and his cast had for the rest of the evening. Besides Hope there was Frances Langford who probably would have drawn the boys out without the rest of the cast. Tony Romano slipped in some soothing guitar music to counter Miss Langford's swingy tunes.

Vera Vague Steals Show

Then came the fun. First there was Vera Vague. Her slanderous cracks about herself seemed very strange to the boys who were seeing her in person for the first time. She is far from the Brenda and Cobina type that she makes herself out on the radio.

Then came the pay off. The rookies of the Second World War were taken back to the Spanish American fracas as a pair of handle bars crossed the stage under the nose of one Jerry Colona. Those men whose sides were still intact went the limit with their laughs as he sang in his own inimitable way, bellowing out the first eight or ten words in one big yell then settling down to real nonsense.

It was a contented crowd of soldiers — that is, paratroopers — who filed back to their bunks. Yes, they rose at 5:45, but it didn't matter that they were up til 12 the night before.

After the show, Hope was made honorary member of the officer's mess and presented with a leather cigarette case."

Letter to Dad from Harland "Bud" Curtis

Postmarked Thursday May 8, 1943
Company F
Camp Toccoa, Georgia

Dear Dad,

Well here it is Thursday night; we are going on a 15 mile march tomorrow morning to a lake somewhere, and where are going to stay there Friday night and come back Saturday afternoon. Gee I don't know how it's going to turn out as it is lightening, but we are still going rain or shine. We'll sleep in pup tents. We've been learning how to put them up, and how to roll our pack; I'm going to carry a walkie talkie for our company commander, isn't that swell. He sent me down to see the officer in charge of the communications section the other day and I have a good chance to go to a radio school here, but anyway in the meantime I'm the only one who knows anything about radios in our company, so I carry this walkie talkie, and I'm not sure what I will do, but I hope I get to use it. I'll let you know after we get back how it turned out. Gosh, Bert could sure get places in this outfit as far as radio is concerned, but there is more to this outfit than radio that he would have to do. The 3rd Battalion hasn't even started to fill up yet, and it won't for about 2 months so if Bert thinks he can make it, he wouldn't be too far behind me. The 2nd Battalion will be filled up in 2 or 3 weeks though, and then we start our basic training.

Bob Hope, and all his gang were here last night and put on a show. It was pretty good (Bud remembered seeing Bob Hope and Jerry Colonia doing a skit. Bob said to Jerry, "I am near an idiot, where Jerry said, Okay I'll move away). *The only thing I don't like about this outfit is that they won't let you have a pass until 7 weeks after basic training, as they want to keep us healthy, because the training is rough. I think by that time we will be shipped to Fort Mackall in South Carolina; so I won't see much of Georgia until I get to Fort Benning for jump training. I sure wish you could see all the trees and mountains around here. There is a lot of hillbillies around here. We see them when we go on runs through the mountains. It is too bad there is an age limit in this outfit as I would talk you into joining up. I'll bet you could whiz right through this training. I think if you have had pervious Army experience that they have to let you in if you are physically fit, so if you ever decide you want to end the war quick, join the paratroops, then Hitler had better start running. If I am not careful I'll be talking the whole family into this outfit. I got the package, thanks I sure needed socks.*

I sure feel sorry for civilians, not being able to get anything without ration cards. How is the gas situation? I sure wish I could come home though, even if things are rationed. It gets pretty lonesome at times not being able to go anywhere like I use to. How is everything in Long Beach? I guess everyone is starting to go to the beach; I sure wish I could lie on a beach again. I guess it is like you always said, you don't appreciate home until you get away from it.

Well I haven't much more to say so I'll stop now. Thanks for your letter, and write again soon; see you then

Love Bud   P.S. Tell everyone hello for me.

Bud was given responsibilty to operate the walkie talkie for Major Seitz

As depicted in the movie "Band of Brothers" men ran up and down Currahee Mountain. All new men to the paratroopers made that run up and down Currahee to see if they had what it took to be a paratrooper. Men who could not make the run up and down Currahee were washed out of the paratroopers.

To help the men keep their minds off of running the sergeants would have the men sing songs. This made the running a little more bearable, and it built unity among the men, a sense of invincibility began to occur. Bud remembered the songs they sang. The most favorite one was one they sang to the tune of the "Battle Hymn of the Republic." It was a thundering sound to listen to the men singing "Blood on the Risers" a tale of a rookie paratrooper who dies when his parachutes doesn't open. It goes something like this:

BLOOD ON THE RISERS

The riser swung around his neck, connectors cracked his dome
Suspension lines were tied in knots around his skinny bones
The canopy became his shroud; he hurtled to the ground
AND HE AINT GONNA JUMP NO MORE!
Gory gory what a helluva way to die
Gory gory what a helluva way to die
Gory gory what a helluva way to die
AND HE AINT GONNA JUMP NO MORE!

He hit the ground, the sound was "splat," his blood went spurting high
His comrades then were heard to say,
A HELLUVA WAY TO DIE
Gory gory what a helluva way to die

> Gory gory what a helluva way to die
> Gory gory what a helluva way to die
>
> He lay there rolling round in the welter of his gore
> AND HE AINT GONNA JUMP NO MORE!
> Gory gory what a helluva way to die
> Gory gory what a helluva way to die
> Gory gory what a helluva way to die
>
> There was blood upon the risers, there were brains upon the chute
> Intestines were a'dangling from his paratrooper suit
> He was a mess, they picked him up, and poured him from his boots
> AND HE AINT GONNA JUMP NO MORE!
> Gory gory what a helluva way to die
> Gory gory what a helluva way to die
> Gory gory what a helluva way to die

Letter to Mom from Harland "Bud" Curtis

Sometime in 1943
Camp Toccoa, Georgia

Dear Mom,

I got your letter today and am glad to hear everything is okay. Well I'm still here and believe it or not, I'm still alive, but I was wondering just how long I would last tonight. Gee whiz we took the hardest run since I've been here, in fact the hardest one in my life. It was up the Currahee Mountain. The steepest part is 4 miles from our barracks here; well we started out from our barracks, and ran all the way out there, and what I mean it's quite a jaunt the way these mountain roads are; well naturally I had my second wind and was feeling great by the time we got to the bottom. That's the way it is; you run for a ways and your really tired; then all of a sudden you get a second wind and you can run all day; but when we started up that Currahee mountain which is about ¾ mile to the very top its almost straight up. Well, I took it for about a ¼ of a mile, and boy you will never realize how out of breath one person can be. They won't run us up the last half mile, because it is just too much too take right now, but we will soon. Anyway, I knew it couldn't be much more to go, and everyone, even the lieutenant was just screaming for air. That's the way it sounds anyway. Just when I thought I couldn't go another step one guy passes out, and three others just stopped. Boy was that bad for my morale. I was just about on the ground myself in the next block we ran. In fact things were really going black. Then the lieutenant said quick time march, that's a plenty fast walk, but boy it sure saved me, and everyone else too. Another ten steps and everyone would have passed out. Well we walked up the last ½ mile; and that ½ mile was the toughest 6 blocks I ever went. It sure was pretty when we did get up to the top. Every direction you looked, there were trees as far as you could see. Sometime I'll get a camera, and take a picture up there and send it to you.

How did those pictures come out that Dean Taylor and I took, send some if they came out alright. Tell Bert (Bud's older brother) to think twice before he gets in any service like this one. Well look at me, I've been here 3 weeks and I haven't even started basic training yet, and it seems like I've been here 3 years; the way they go about toughening you up. It wouldn't make any difference if a person dropped out of one of these runs now or not; oh it would but what I mean is they won't kick you out, but when you start basic, you can't drop out of anything. Well I said things would get tougher, and they sure have, but I'm determined to stick it out, as I know I'm as good as any of these other guys. I've found that out for sure, so nothing can stop me now, unless I drop dead on that Currahee Mountain and what I saw of it today, it wouldn't be hard. Tell Dad I sure appreciated his letter and I'll write to him soon, but this letter will have to do for now as lights are about to go out.

Love Bud

P.S. Thanks for the cookies they were swell

P.S. I tried to call you up this Sunday, but I couldn't get the call through. I will write soon.

Well I didn't mail the letter last night so I'll write a little more; it is lunch time right now, so I've got a little time. I saw a lieutenant of the communications outfit, and I'm pretty sure that I'll be transferred to Headquarters Company some time soon, or maybe I'll have to wait until 7 weeks after I start basic training. I'm not sure, but I have got a good chance to be a radio operator. I'll be in the same Regiment and Battalion, but just a different company and instead of carrying a machine gun and such, I'll carry a radio. Well I don't know yet how it is going to turn out so I'll write more later. I had better close now and mail this as I haven't much time till I fall out for inspection. Write soon.

Tell everyone hello   Love   Bud

Letter to Folks from Harland "Bud" Curtis

Postmarked May 19, 1943
Company F, Camp Toccoa, Georgia
Monday

Dear Folks,

Well here is another short letter. It is after 12:00 am and I have to get up at 5 and then we are going out to the firing range about 10 miles out. We are going to go with full pack. I guess I will carry the walkie talkie again. We are going to be out there until Saturday so I guess I won't get a chance to write to you, but if it is possible I will, and let you know what kind of a shot I am.

I don't think you had better send a camera as they have been restricted just recently. So I can't use it anyway. How is everything at home?

Boy have I been rushing around tonight. I had to clean my machine gun and then get all my equipment packed for this march. Boy I sure like that machine gun, but I won't get to use it for a while yet. Boy in this outfit you have to know all about every weapon. How it works, all the parts and how to use them.

Well 3 weeks now and we will be in basic. I mean those that pass on shooting. Those that don't will get kicked out. So I'll let you know as soon as I can if I made it. I don't see why I can't though. I sure hope it doesn't rain as we will be camping out in our pup tents. It is more fun, but I like to froze to death on the last outing.

Well I am so darn sleepy I had better hit the hay. So I'll thank you again for all the nice things you have sent me and I will write as soon as I can.

Love

Bud

In the book "Battling Buzzards" author Gerald Astor reported, "The troopers of the 517th qualified on target ranges with their assigned weapons, Colonel Walsh initially demanded that every man achieve the top designation of expert with his piece, and in keeping with his belief that all should be familiar with other armament, required the next highest rating of sharpshooter with a second weapon and the minimum of marksman with the crew served guns and mortars" (Astor, 1993).

Sometime in June 1943, all units were notified there would be a camp wide required physical fitness test. Those who could not pass would be processed out of the paratroopers. Bud and the other men of the 517th fell out around 1500 hours (3:00 pm) for the test. Bud was standing over by the pull up bars when he decided to perform some acrobatic stunts for his friends. Bud jumped up and grabbed the bar and pulled himself straight up to where half his body was now above the bar with his arms straight holding himself up. This quickly caught the attention of many of the troopers. In fact his commanding officer, Captain McKinley, called out to the other men of the company to assembly around the pull up bar, and watch Curtis perform some acrobatics. Bud started to perform by swinging around the bar. The men were astonished. As Bud was swinging around he quickly changed his hand position and now he was swinging in the opposite direction. Then he swung up above the bar to where his legs were over the bar and began swinging around the bar with his knees. For the grand finale he swung under the bar letting go with his knees and quickly turning his body around midair grabbing the bar with his hands swinging back up to sit on top of the bar. The men went wild, cheering and applauding. The Captain shouted out that's the end of the show fall in for the physical fitness test. Oh, by the way Bud past the physical fitness test with flying colors.

Letter to Mom from Harland "Bud" Curtis

Company F, 2nd Battalion
517th PIF
Camp Toccoa, Georgia
Postmarked May 21, 1943

Dear Mom,

Well it is Friday, 12:45 pm; we've got to fall out for drill in 10 minutes so I'll finish this later. I just got your letter this morning. I am glad everything is going along alright. Well I guess I'm going to be here for a little while as we won't even start our basic training for another month because it will take that long for this Battalion to fill up, so in the meantime were just getting into condition. We get up at 5:30 A.M. and go on a two mile run before breakfast. This is to wake us up. Then after breakfast we have an hour of calisthenics, then during the day we drill or have different. Got to go now.

Hello again here it is 8:30 P.M. To continue, we have drill or classes or different things, and at 4:00 P.M. we go on a run from 5 to 8 miles (more fun). Boy the sun is sure hot and there we are running up mountains. We are getting into condition so that we can run 60 minutes without stopping. Heck I can't even get a pass to go to town for about 3 months. We have to wait until 7 weeks after we start into basic training, and we won't start our basic for a month, so all I have to do for the next 3 months is to get healthy.

The food is pretty good now. I was at another Battalion when I wrote that last letter, and I was eating at this mess hall where these fellows first come to, and they feed you lousy, just to discourage you; they figured they wanted you to quit right then if you are going to, rather than waiting until they waste good time on you. I'm in this better Battalion now and things are pretty nice.

I sure wish I was home to build up the battery for you as things get pretty boring around here at times. How is everyone? Gosh if Bert has 20-40 eyes he could get in here. He would get a lot farther than I would with his education. What he should do if he is expecting to get in the service soon is to join the Army and volunteer for the paratroops. The main thing to get in this outfit is not to ever had any broken bones or even if you have don't tell them.

If he joined the Army right now and volunteer for the paratroops, he would get sent right here within 2 weeks, and we would both start basic training together and would probably be together all the way through. Even if he didn't pass the physical he could always go to an officer's candidate school. And if he did get in here he would be a cinch to make staff sergeant in this Company's radio school, and he would be the Company Commander's right hand man. That is what I'm trying for, but I haven't got as much chance as he would have. Don't let Bert get all enthused about this, if he has a chance to stay out much long as it is a rough life, and plenty tough. It sure would be swell if he would get in the same outfit with me. But he would have to join right away if he did want to get in this outfit with me as in a month we will be staring basic training, and if he wasn't in by then he would have to wait for about 2 months more in the 3rd Battalion until that filled up, and then he would always be behind me. Tell him not to join the Navy, as he will have a heck of a lot better chance to get somewhere in the Army. I know he is physically fit all but for his eyes, but have him find out how they are for sure. This is a long way to come just to find out your not physically fit. But as I said if he didn't make it he could always go to O.C.S. (officer candidate school) and really get places. He would be a 90 day wonder and I would have to salute him. Well write back soon and tell me what he thinks about this, and I'll explain the exact way to go about getting into this outfit, but if he can stay out of this war all the better.

Well it is getting about time for lights out so I'll have to hurry. How is Glen? (Gen Dawson, Bud's cousin) I'll bet he had a rough time, but I'll bet if things go as I expect I'll have a much rougher time. (The bugle is blowing for lights out so I'll finish tomorrow)
Goodnight

Well here it is Saturday morning, gosh we have got a big inspection for today. I'm sure going to make a good boot black. I've sure shinned a lot of shoes since I've been here. Well I've got to go now. Oh send some airmail stamps, and some white union socks and send me a little pen light. One would come in handy around here. Tell Dad thanks for writing me.

By now
Love Bud

Bud came from a very close family. Bert was his only brother and they were close growing up. Bud must have missed him terribly and thought if Bert could get into the paratroops they could be together.

Letter to Folks from Harland "Bud" Curtis

Tuesday Sometime in May 1943
Company "F" Second Battalion
Camp Toccoa, Georgia

Dear Folks,

Well right now I am sitting in my pup tent out on the range. What I mean I am roughing it. It has been raining for the last 2 days. You should have seen me coming out here I march 5 miles without stopping with full pack; a machine gun on my shoulder, and a walkie talkie on the other, and it was raining cats and dogs all the way. This life is sure rough.

Everyone is running around in their shorts dripping wet. It is kind of hard to get your close dry in the rain. We will be here for about a week or more. The Major said everyone who qualifies on the rifle will get passes to go to town.

Well I have already qualified so I'm all set. I sure was glad to hear that. I was afraid I'd have to wait 7 weeks as I said before.

Well it is getting dark and I am getting cold. I will write to you soon as I can if I don't catch double pneumonia in the mountains. Don't worry, I won't. They won't let you get sick in this outfit. You are too busy.

Tell everyone hello.
Love Bud

In this letter Bud talks about the grueling running and force marches they were required to perform. On June 28, 2007, I was at the Chicago airport waiting to fly to Washington, D.C. to attend the 517th reunion. There I was fortunate to meet with a 517th trooper, John Anderson, of "F" Company. With his son Bill Anderson, he told us of a day when LT Lissner took them on a 20 mile force march. John Anderson, said, "We wore our helmet, pack, and carried our rifle. We were instructed to have NO water in our canteens. While on the march about 10 or 12 miles a trooper named Fred Winke was drinking water from his canteen. LT Lissner quickly told him to stop drinking the water, and to pour it out of his canteen. When we got back to camp PVT Winke was processed out of the 517th. I never saw him again."

Letter to Mom and Family from Harland "Bud" Curtis

Company "F" Second Battalion
Camp Toccoa, Georgia
Postmarked May 26, 1943

Dear Mom and Family,

I got your letter today, and I'm glad to here that your feeling better. I've been getting the press regularly now; it started dated March 17th. I've gotten all your packages. Thanks a lot for the cookies, socks, flashlight, etc. Dean Hildreth has volunteered for the paratroops and he is at Fort Benning, Georgia getting jump training; he already had basic training. He will get his wings before I do, but I'm getting better training the way I'm going about it. There are 4 stages, each a week long (A, B, C, D) stages. A consists of physical hardening, and they try to join all the rough training like running and such into that one week and boy it is plenty rough. I won't have to take A stage because the fellows who have come out of this camp practically laugh at it because it is so easy for them; so when I do get to Fort Benning I'll go into B stage; anyway it will only take me 3 weeks to get my wings (when I do get to Fort Benning) so you see the parachute training is really the easiest part of my whole training. It is the most dangerous part and you have to be fit pretty good to do it, but it is what I am doing now that's hard to take, but don't worry about that. The fighting I'll do if I do go across will be either desert or jungle fighting. We've been shown quite a few Army restricted films of the fighting we will have to do, and boy I'm not particularly anxious for it, but I'll take it as it comes.

I sure had fun on that hike; we went up to this lake and was it pretty. There is nothing like the Army when you go on a camping trip. The best part of it was during the march to and from the lake. I walked side by side with our Major at the very head of the Battalion and I carried a walkie talkie, and gave orders to all the companies. I sure felt important. Our Major is only 30 years old and is he rugged, he set the pace, and we walked at 130 steps per minute; that's about 4 miles per hour (with full packs). We walked up a 45 degree hill for a mile in 10 minutes, and boy everyone was really tired out. They know just about how much you can take. We took a 15 minute rest period as soon as we got to the top, and the way they do it is the officers have to keep standing up while all of the other men sit down and rest. So you see the officers in this outfit sure have to be able to take it, you couldn't go anywhere to find a better outfit. It was really interesting talking to the Major all the way; he told me all about his paratroop invasion of Tunisia.

I'm pretty sure I'll be able to go to a radio school as the Company Commander seemed quite pleased at the way I handled this outing; and he said he would give me a reference. So as far as I know now my last 6 weeks of basic training, I'll go to this school. Guess what? I've got to go on K.P. (kitchen police scrubbing pots and pans) tomorrow. They go down the list alphabetically, and it is my turn. Oh well I won't get it again for a while unless I do something wrong. It has been raining all day. It seems to rain regularly every other week;

it really gets sloppy around here when it rains, and it really gets hot when the sun shines. What a place.

Well we will start basic training pretty soon now as the Battalion is just about filled. The Major (Major Seitz, 2nd BN CDR) told me out of 3000 guys that passed the physical so far only 250 have been let into the outfit. He interviews everyone of them, and he is plenty particular I guess. That's why it is taken so long for us to get started.

Well it is bed time again and I've got K.P. tomorrow so I had better get some sleep. Write soon and thanks again for everything. Oh I got grandma's letter and I'll write to her probably tomorrow.

Tell everyone hello, see you soon   Love Bud

At the 517th PRCT reunion in Washington D.C., on July 1, 2007, Bud spoke with now General Seitz about how he and the other men were accepted into 2nd Battalion of the 517th. General Seitz said, "I interviewed 3,100 men and only allowed 700 into the 2nd Battalion. Headquarters was putting a lot of pressure on me to get the battalion staffed, but I was looking for men with spirit and heart, not muscles." One of 5 men became the best of the best, which was only 22% of those who were trying to become paratroopers. These were truly American's finest fighting men thanks to General Seitz's determination to have nothing but the best.

Letter to Mom from Harland "Bud" Curtis

Company "F", Second Battalion
Camp Toccoa, Georgia
May 30, 1943

Dear Mom,

I got your letter with the pictures in it yesterday. I haven't much time to write this letter so I'm going to have to hurry.

I've had a heck of a cold the last couple of days; if you can send some nose drops; they haven't much to give you for a cold except an aspirin. So if you can send some, swell.

If you can't get a tube for the radio it is okay, because another fellow has his radio here now. This Army life will be a lot better when I can start getting passes to go to town. It won't be long now until we start basic training.

Dean says he is sure getting worked over, there at Fort Benning. They are really on you the first week your there. (that's A stage). I won't have to take that at Benning, because I'm getting it all the time I'm here, but Dean is getting it the hard way.

So Willard Hill was shipped across already. I really haven't much to write about right now.

I'll bet Jill is glad to be home.

Thanks for everything you sent me; I sure need them. You can send me a wash cloth if you have time. Thanks again.

Write soon.   Love, Bud

In addition to qualifying with the rifle Bud was trained to fire the M-9A1, 2.36" rocket launcher commonly called the "Bazooka." This new folding bazooka was an immediate success with Airborne troops and was easy to jump with because it could be broken down into two section.

Letter to Folks from Harland "Bud" Curtis

Camp Toccoa, Georgia
May 30, 1943
Sunday

Dear Folks,

Here I am on guard duty. I've been on since 11:00 A.M. this morning and I'll get off at 7:00 P.M. Monday. You walk post for 2 hours then your off four hours. It is not bad, but you don't get much sleep. It is 11:00 P.M. I don't go back until 1:30 A.M. You ought to see it rain; it sure has rained a lot around here lately.

Boy we sure had excitement today. The whole first battalion got food poisoning today, and there were guys passed out and laying all over the camp. They were really sick. It sure kept the medics rushing around with their stretchers and ambulances. I felt sorry for the ones that weren't quite so bad off. They had to run through drills until they about dropped so they could sweat out the poison. I'm in the day room right now, and it is about to close up so I had better cut this short. I didn't have anything to say anyway. I just have plenty of time to do nothing until I go on guard. I guess I should get a couple of hours sleep.

I told you I had a pretty bad cold. Well I went down to the dispensary and they gave me some pills and this cold sure went away fast. I wonder what was in them. I could still use those nose drops I guess; so whenever you get time, okay, but there is no hurry. I was going to ask you to send something else but I can't think what it was. Wasn't important I guess.

Well I'm being run out of here so I'll write again soon.

Goodbye now - Love as always Bud

Letter to Everyone from Harland "Bud" Curtis

Company "F", Second Battalion
Camp Toccoa, Georgia
Postmarked June 7, Saturday 1943

Hello Everyone,

Here it is Saturday. Time sure is going by. We have had a swell time today. We had a field day all day. We played games went over the obstacle course. Played baseball. Well just about everything. Boy it sure is hot here. I am up on the roof getting a tan. They think I'm crazy around here because it has been so hot, and here I am. I took 3 showers today and I'm going to take another one in a little while. The Major told us today that we won't start basic training for 4 weeks; so I'm going to be here one month more than I thought before I can get a pass. Well you will know where I'm at anyway.

I got a letter from Garth, he is in San Diego; he's in the Navy.

Guess what, our barrack is quarantined for the measles. We can do everything, but go to the PX and the show. And that's all there is to do, so were just out of luck. I shouldn't tell you this because it's not suppose to get out, but on our run yesterday we passed a fellow laying all bent up at the side of the road; he was really moaning. He was from another Company and they had left him there while they went back and got the ambulance; well in the meantime we came along and stopped. The ambulance came up then, and while the doctor was looking him over he died. Well that's an example of how tough this outfit really is. Somebody will get plenty of hell for his dying, but that won't help him now. What I mean is when a guy runs so darn far he drops dead; that's going a little too far. Don't worry about me, because I can take it all right, and so can the rest of the fellows, but sometimes a guy gets in here that shouldn't be, and that's what happens.

They are picking about 35 fellows from our company to go on an endurance test, and I'm one of them. Boy I'm really going to get worked over now. They'll get us up at any hour and we'll march – (oh just use your imagination on the meaning of endurance test, and that's just what it means) they want to find out first how much we can take. So I'm going to be a Ginny pig.

This is really a swell outfit though; I wouldn't get out of it if everyone dropped dead. I just wish I could go to town, or do some things like I use to could.

What's been going on in Long Beach. Send me my bathing trunks, and a wash cloth will you. I can run around in my bathing suit any time I want on my own time. Then I can get suntanned all over. I'm really brown from my waste up. I'm beginning to look a little bit rugged and healthy. You will never know me when I come home in those boots and wings and looking healthy.

It sounds like the 4th of July around here. The 1st Battalion is out on the rifle range, and there throwing hand grenades, shooting machine guns etc.

It doesn't get dark here until about 8:30 pm. It will probably rain pretty soon. It always does, all though you could never tell by looking. The sun is out bright, and its 7:00 o'clock.

The thing I like about the paratroops is most of the fellows here are just 18 to 20 years old (that is about 80%) the other 20% range from 20 to 30 years old. It is just like going to school, except school was never like this.

Everybody that goes by looks up at me here on the roof, and go away mumbling things. I guess they think I'm a little off the beam.

Well I'm going down and take that shower now, so I'll leave now.

Write soon.

Love Bud

Letter to Folks from Harland "Bud" Curtis

Company "F" Second Battalion
Camp Toccoa, Georgia, Postmarked June 14, 1943
Friday

Dear Folks,

I just got your post card a few minutes ago. No, I haven't written this last week that's why you haven't got any letters. I've been so darn busy this last week that I just haven't had time. We have been practicing positions and aiming with rifles all this week, and then at night I go to a special machine gun school; so that takes up all the evenings. Monday, Tuesday, Thursday and Friday only; 4 nights a week. We will be out on the range firing all next week. We will camp out in our pup tents so I guess I won't be able to write next week either so don't worry. It is just about time to go to school so I'll have to hurry.

I sure did like those cookies. I wish I could be home to have a big glass of milk to go with it. That sure was a swell box Dad built for the radio. We will be half way through basic training before we even start, at the rate we are going now. The things we are getting are the same as in basic, but they have to have something for us to do. Well I'll sure be glad when I get through with all of this. We ran with rifles and helmets yesterday, boy that was rough.

If Bert can stay in the ship yards that is the best place for him I guess. I would sure like to be just working 10 or 12 hours a day. When I signed the papers for this outfit, they said I would be on duty 24 hours a day, and I guess they weren't kidding as I haven't even got time to write on weekends. Sunday is our day off. Well I really have to be going. Don't *worry if I don't write next week as I'll be pretty busy. Write often and thanks again for everything. Love, Bud*

P.S. I'll write Saturday or Sunday if I have any time to myself. After 7 weeks of basic I'll be transferred to Headquarters Company and go to a special radio school for 6 weeks.

Letter to Folks from Harland "Bud" Curtis

Company "F" Second Battalion
Camp Toccoa, Georgia, Written June 28, 1943

Dear Folks,

It was sure nice to hear your voice again. I was at the drill field when I called. I was down there instructing the fellows that haven't fired yet. I don't know if you got my other letter, but I have already fired and I got one of the highest scores. They gave me an expert Medal and the freedom of the Post area. Now I don't have to march to the PX or the show.

Well anyway they made me an instructor, and I went over to a phone they had close by the drill field where I am teaching these guys. It only took 10 minutes to get the call through. I just took a chance. So Bert got married. Give him my best wishes and such, but what is he going to do if he goes into the service? I think it would be better if he stayed right there in the shipyards if he can. How long is he going to be in Sequoia? I didn't quite get when they were married. Write and let me know how everything is.

The fellows that I have been instructing are going out on the rifle range tomorrow and I will have to go with them. We will camp out there for about a week or so. I hope it doesn't rain, but the way it is coming down right now, I wouldn't place any bets. I think I will get to shoot my machine gun when I am out there. I sure hope so.

As soon as we come back, we will start our basic training. It is about time. I have been here for 9 weeks today. As soon as I complete seven weeks of basic I will be transferred into Regimental Headquarters and take a 6 week special radio school. Also after 7 weeks I can get passes that is what I am going to like. Time will pass by pretty fast now I guess. I probably won't be able to write as much as I would like to as they really keep you going in basic training. They will have us going day and night, 7 days a week.

Well I have to hurry now and clean my rifle for inspection. It is 6:00 A.M. Monday morning. I will write to Bert as soon as I can.

Bye now, Love, Bud

Letter to Folks from Harland "Bud" Curtis

Company "F" Second Battalion
Camp Toccoa, Georgia, Postmarked July 1, 1943

Dear Folks,

Here it is Sunday night, we came back from the range last Saturday (yesterday). You should have seen me. I was covered with mud from head to toe. It rained everyday we were there, it didn't hardly let up a minute. I was soaking wet all the time we were out there. I had a lot of fun firing my machine gun though. I must be getting rugged. I didn't have so much as a sniffle of a cold.

Send Bob Douglas and Gary H. addresses. I have lost both of them. I have moved around a lot the last 2 weeks. Note the new address. I am here at Regimental Headquarters now. We start basic training tomorrow. I have had most everything we get in basic already, but I will just have to take it again. We will take 7 weeks of basic, then the fellows here in Regimental Headquarters will go 6 weeks to different special schools. I will go to a radio school. There is a small possibility that I might make my jumps at the end of 7 weeks and then take 6 weeks of school, but I doubt it.

It is about time we got going. I will be here 10 weeks tomorrow. Next weekend I get a pass to go to town.

Well it is after eleven and I have a big day tomorrow. Wish Bert good luck. Write soon.

Love, Bud

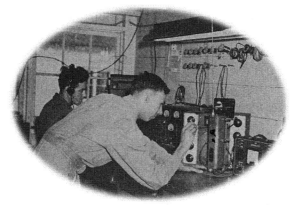

Bud's Company Commander 1st Lieutenant McKinley of "F" Company discovered Bud's ability to work and understand radios. An opening occurred in Regimental Headquarters and Lieutenant McKinley had Bud transferred. Bud was very upset being transferred out of "F" Company. All of his friends were there; however, the transfer may have saved his life. He was now out of a line company and into a technical job in Regimental Headquarters Company. His job would not be as an infantry private to fight, but now to keep the radios working and communication between the Regiment and it's Battalions.

These are Bud's weapons qualification badges. There are three levels of weapons qualification badges; Expert, the highest qualification, Sharpshooter, the second highest, and Marksman the lowest. On the left is Bud's Expert badge in Rife (M-1), Machine Gun and Pistol. On the right is his Sharpshooter badge in Mortar, Hand Grenade, and Carbine Rife. He qualified and instructed marksmanship at Camp Toccoa, GA, in 1943.

Chapter 4

TRANSFER TO REGIMENTAL HEADQUARTERS, CAMP TOCCOA

November From the Commanding General, "All known physical test records have been broken by your organization!"

    Although, it will never be known for sure, Bud's transfer to Regimental Headquarters probably saved his life. Bud displayed expert marksmanship and was an excellent infantryman; however, his company commander saw more in him. Bud had an ability to think on his feet, and his knowledge about radios was more than other men. Thank goodness he had a brother who was an expert with radios, and would include Bud in the interagency of how they worked. Both boys would build radio sets when they were kids. Taking a radio class at Poly High School also gave him the technical training his "F" Company Commander would recognize. Lieutenant McKinley made the decision that this man could contribute more than being a line infantryman. Bud's radio skills were needed in Regimental Headquarters and he was quickly transferred. There he met his new section sergeant; Staff Sergeant (SSG) Pease, who in later times Bud would introduce to a girl in the town just outside of Fort Bragg, N.C. They hit it off so well they were later married.

Letter to Folks from Harland "Bud" Curtis, Sometime in 1943

Regimental Headquarters, Camp Toccoa, Georgia
Monday

Dear Folks,

  Well here it is Monday, and I'm on a break right now so I thought I would write you a letter. You'll have to pardon this paper and pencil. This is all I have right here.

  Well I'll tell you what I have been doing. We went out to the firing range early last Tuesday morning and we stayed there until Saturday. We fired from the time the sun came up until it went down. Well I don't know how I did it, but I came out one of the highest scores in the whole Battalion. There were only about 20 that made "Expert" (the highest marksmanship rating) and most were officers. There is marksman, sharpshooter, and expert. Well I seem to be doing alright for myself; I am an instructor for the other barracks that hasn't had this yet. I am on the drill field yelling at these dumb dopes. I'm as mean as any sergeant you can find. We still haven't started basic yet, but what we're doing now is what you get in basic. I won't have to do anymore firing now. Only the ones that flunked (boloed) and the new guys will fire, but I have to go along and be an instructor. So I guess next week I'll be out sleeping on the ground again. Boy there was one day we had one of those Georgia storms when we were out there and what I mean we were wet and cold, but that didn't keep us from firing. The M-1 rifle is sure swell. I can really cut up that bull's eye. The very first practice shot like to scared me to death though. After that I did all right. Another thing I like is that I get to march at the head of the Battalion with the officers with that walkie talkie. I am assigned to Regimental headquarters and I will be transferred after 7 weeks after basic. I like the Company I am in, but I would rather be in communications and radio though. I have been here eight weeks today and I sure am tired of not being able to go to town. There is nothing I can do about it though, time will pass a lot faster when we start basic. If I don't write to you as often as I should don't worry about me as I have a good reason. I'll be out on the range next week again so I won't be able to write there either, but they bring our mail out to us.

  Well all I have to do is stick with it and I think I can get somewhere in this outfit. I'm going to a machine gun school, and when I get through I will get to fire it. I'll be an instructor in that too. Then I want to go to a mortar school. I'm going to know how to operate everything the paratroopers have when I get through, and I'll be a radio man too if things go right. I'm going to get a pretty medal for being such a good shot. It sure did surprise me. I got a higher score than most of the officers. Dean Hildreth (friend from high school) has made his jumps at Benning now. I sure wish I was there with him, but I think I can get father and better training the way I'm doing here than he will. Well I really haven't any more to say so I'll leave now.

  Love as always

  Bud

P.S. I got your package today with trunks and cookies. I just got a chance to open it and we had to fall out. Thanks a lot

  In this letter Bud relates his excitement about receiving his Paratrooper Jump Boots. These were very special to paratroopers. Ruggero (2003) related "Troopers kept their boots highly polished, and to better show them off, they began to tuck the bottoms of their trousers into the boot tops, a practice they called 'blousing. The regalia was eye catching, and for a lot of young men putting on a uniform for the first time, the attraction was no small thing."

Bud in later years told the story that he was not allowed to blouse his new paratrooper boots because he had not completed jump school. He said, "When we went to town we bloused our boots anyway and when talking with the young ladies of Toccoa, we held our right hand over our left breast pocket to hide the fact that we didn't have jump wings yet."

Letter to Mom and family from Harland "Bud" Curtis

Regimental Headquarters Company
Camp Toccoa, Georgia
July 26, 1943

Dear Mom and Family,

    Here it is Monday night and I guess It is about time I wrote to you all. I was in Atlanta this weekend; I tired to call you up but there was too much of a delay. We finally got our boots, boy do they look nice. They let us have them now so we can break them in. The troopers sure look nice when they're all dressed up. They stand out among all the services when you see a crowd of servicemen on the street.

    I sure am tired. I got in at 5:00 A.M. this morning and had to fallout at 6:00 A. M. Oh boy! I use to be able to do that when I was a civilian, but it just about threw me today, especially on that darn run this evening.

    We will be leaving here between the 4th and the 10th of August. I guess I will go to Atlanta this weekend. It will be the last time I will be able to go until I get to Fort Benning. Atlanta is really a nice place. It reminds me of San Francisco except there aren't any hills. I never saw so many WAACS (Women's Auxiliary Army, Corps Service), WAVES (Women Naval Service), WAMS (Women's Auxiliary Marine Service), etc. etc (these acronyms referred to women in the different services). It is a regular service city.

    I have been on a regular vacation since I have been in Regimental Headquarters. We don't have so many inspections and we can get weekend passes and the other companies can't. I am broke but it was worth it to be able to go someplace. I won't send you a picture until I have my wings; then I will have a full length picture taken.

    As far as I know now when I get my 10 day furlough I won't get any traveling time and they say the ones from the west coast had better forget about going home. There are a lot of fellows from California here, so I think they are going to have to make allowances for traveling time as a lot will just go any way. But don't worry about that until the time comes.

    How about having the Press transferred to this address I am at now (meaning the Press Telegram Newspaper). Well I am going to bed now. I will write as often as I can. Tell everyone hello,   Love, Bud

These are Paratroopers "Jump Boots", M-1942. No other soldiers were allowed to wear these boots except paratroopers. Bud was very proud of the elite status these boots gave him as one of the few who could be called a paratrooper.

Letter to Folks from Harland "Bud" Curtis

Regimental Headquarters Company
Camp Toccoa, Georgia
Dated August 2, 1943

Dear Folks,

It was sure nice to hear your voice again, but you didn't seem very glad to hear from me or you were pretty sleepy. I won't call up any more though; it to expensive anyway. I'll pay you back as soon as I can get paid again. It won't be long now until I get jump pay. I am starting radio school tomorrow. They cut our basic training off short because we had already been here so long.

I am going to send my watch home and I want you to sell it to help pay back some of this money. I know you need it. I wouldn't ask you for it except I owe out quite a bit. Some rat got a hold of a pass key to the foot lockers in our barracks and stole every body's money including mine. He went over the hill and the finally caught him in Chicago. Well he won't be doing anything for a long time, but that doesn't help me or any of the other fellows out, because he had already spent it all. I can borrow money from the fellows here, but it is not so easy to pay it back when they need it.

Well it is a sure thing now that it won't be over 6 weeks until I start jump training, because this radio school only lasts 6 weeks. I will sure be glad to make my jumps and get my wings. I already have my boots, and do they look swell, and it is also going to be nice getting a $100.00 instead of $50.00. There are so many things taken out of that $50.00 anyway that's all I ever see of it. It is not much over $30.00.

I am glad Bert got his deferment; I hope he never has to get into this. You had better appreciate him being so close around as I doubt if you will see me for a long time. I will get a 10 day furlough after my jumps, but I can't get any traveling time, so like the rest of the fellows from the west coast I will just have to take it as it comes.

It won't be much longer after we have made our jumps until we get sent across. I can promise you one thing for sure, and that is I won't be in the country for Christmas. They tell us they have our sailing date set already. Of course none of us know where it is, but all you have to do is use common sense and figure it out. We won't be any good around here after we have finished all of our training.

I will be in Camp Mackall within the week. It will be nice to see Dean. It is nice to see someone you know again. Dean may have his faults, but he must be okay, or he couldn't be in this outfit. Eddie Hunter (the guy who went with Bud and Garth to San Francisco in 1942) is in the Hawaiian Islands somewhere. He is in the Tank Corps. Dean Taylor is some where over seas. He wrote me a letter and told me he had crossed the equator. He said he really went through a rough initiation (When men cross the equator for the first time the U.S. Navy makes them complete an initiation. The new men are called "pollywogs" and are made to crawl through garbage while other sailors beat and kick them. One of the fattest sailors portrays King Neptune, and each "pollywog" must crawl up to him and kiss his belly with all kinds of slop smeared on it. Once the initiation is completed each man becomes known as a "Shellback.") That must be really something to go through.

Garth is still at San Diego. He is a gunner's mate. I wouldn't trade places with any of them. The signal corps is taking over this camp. There are about 500 of them here now. You can sure see the difference between this outfit and the regular Army. We have had a lot of fights since they have been here. You have to laugh when you see them and these paratroopers together. They look like a bunch of 4 F's (4 F is the designation by the draft board for someone who is disqualified for military service). We are all brown and they *are white as ghosts. They are getting a few ideas from us about physical training. They tried doing a little double time (running) today, and they didn't go more than 200 yards when half of the company had fallen out. Then we go on a 6 mile run and came back hardly sweating. We really don't consider our self as part of the Army. If you ever see any paratroopers and regular Army men together you will know what I am talking about. Oh well they will probably live longer.*

I guess it is getting pretty warm there now isn't it? I wish I could come home for a couple of days and go to the beach again. When you get my change of address call up the Press Telegram (Local newspaper) *and let them know.*

Try and find out what Roger's address will be. And find out when Jill's birthday is and I'll send her a card or something.

Don't worry if I don't write as often as I should. I am kept pretty busy. I guess I will be even busier when I start school. I wish Bert was here to help me out in this radio school.

Well it is way past taps and I have a lot to do tomorrow.

Write whenever you can.

Love Bud

Bud started Radio School in August 1943, learning Morse Code, and other communications techniques

Letter to Mom from Harland "Bud" Curtis

Regimental Headquarters Company
Camp Toccoa, Georgia
Postmarked August 2, 1943

Dear Mom,

I am really sorry I haven't written sooner, but I have really been on the go. Well I won't be here much longer. We are getting shipped to Camp Mackall, North Carolina (Camp Mackall was named after Private Mackall who was killed in action in the first combat parachute jump into Tunisia Africa in 1943. The camp was pronounced McCall but the soldier's name was pronounced Mack-All. His family was always upset that they never pronounced his name correctly.) *We will probably be out of here by the 6th of August. I sure am glad to be going there. Dean Hildreath is there too, so that will be pretty good too. It is nice too see someone you know. Dean said that when he made his fourth jump at Fort Benning there was a 25 mile ground wind and when he came down he hit the ground* twice and was knocked out for 10 minutes. He said there wasn't anyone that walked off that day. A few broke their legs. How about that? Well anyway he is a full-fledged paratrooper and he is at Camp Mackall now. I was in South Carolina last weekend. Did you get the card I sent you? I will be pretty close to Norfolk, Virginia. I might go up and see Pat (Pat Schubb was the daughter of a Naval Officer who was stationed in Long Beach. Bud knew her from his days at Wilson High School. She lived in the neighborhood about two blocks away. Here father was transferred to Norfolk Virginia and the family now lived there. One Friday night Bud took a greyhound bus from Fayetteville, N.C. (located near Camp Mackall) to Raleigh Durham, N.C. Then he took another bus to Norfolk, VA. Bud arrived in Norfolk at 6:00 A M. Sunday morning. He telephoned Pat and told her he was in town. Pat told Bud where she lived and he took a taxi to her house. They visited for about 3 to 4 hours, and then Bud had to leave to catch the bus back to Camp MacKall. Bud said his goodbyes, and took the taxi back to the bus station where he got back on the bus going to Raleigh Durham, N.C. In Raleigh Durham, N.C., Bud went to get on the bus going to Fayetteville, N.C. He was in for a shock when he found the bus was full, and had no more room for another passenger. Well, Bud had to get back to camp; he could not be absent without leave (AWOL). Using his ingenuity Bud climbed through one of the bus windows and crawled into the luggage rack above the seats. The bus was pack with soldiers and this was the only place he could ride. The bus driver never knew he was there. Bud rode in the luggage rack all night and arrived back in Camp about 5:30 A.M. Monday morning. Bud never saw Pat again.)

Boy is it hot today, Wow! We have to fall out for a parade in 15 minutes so I will have to hurry.

Can I send the radio in the box. I am going to be moving around quite a bit for a while and I think it will be best if I send it home.

If I don't have to work this Sunday, I am going to Atlanta. I will drop you a card from there. Well things are moving pretty fast now. It won't be very long until I am up there looking down and wondering what the heck I am doing up there. Boy that is what I'm looking forward to as I will get a furlough pretty soon after my jumps and I will come home with those boots and wings. Won't that be something. Well I have to go now. I will write as soon as I can Tell everyone hello

Love

Bud

Letter to his brother Bert from Harland "Bud" Curtis

Regimental Headquarters Company
Camp Toccoa, Georgia
August 5, 1943

Dear Bert,

Here I am sitting in a radio class trying to catch up on a little sleep, but this guy is worse than Farraund (This was Bud's radio teacher at Poly High School in Long Beach, California). *In fact he has hooked up a microphone and he is making so much noise it makes my teeth rattle.*

I guess it is about time I wrote to you and wished you luck and all that. I am glad to hear you got a deferment. I'm sure glad to hear that because if you can keep out of this the better off you will be.

Damn, just a year ago I was a happy civilian living in San Francisco. That is a good one for the books. There we were Eddie, Garth and I all enjoying life, and now look at us. Here I am thousands of miles away. Garth is in the Navy, and Eddie is in the Hawaiian Islands. How about that?

Well it won't be long now until I am going to be falling through space wondering why my parachute doesn't open. It had better open because I don't like the looks of my blood all over the ground.

I wish you were going to be there with me to push me out the door. No kidding, I can hardly wait to get started in my jump training, and get my wings. Boy, do those boots and wings look good. I sure hope I can come home before I get shipped across. Then I'll just go down town with you and you can see the difference these boots make.

If you ever get a chance to do any traveling be sure and go to Atlanta. It is really a nice place. No kidding there are more good looking girls here than you can imagine. Atlanta reminds me of San Francisco, only there are no hills and apartment houses here.

I sure hope this month goes by fast, because some rat swiped my money and I owe out most of my pay check for the next month from what I have had to borrow. A $100.00 a month is really going to be nice. Well it will be about the first of September when I start jump training. One more month. I guess I can hold out that much longer.

This radio class is better than being outside working my ass off. All I have to do is go on a 5 mile run at 4:30 P.M. every evening to keep in shape.

I will be in Camp Mackall, N.C. about next Tuesday. If Dean Hildreath is there and if he has any money I am going to go to Norfolk, Virginia. I always wanted to see the place. If they don't give me any traveling time on my 10 day furlough I am going to travel all over the east coast. New York, Pennsylvania, and all over. I won't have enough time to get home and back in 10 days; so if they don't give me traveling time I will just have to wait until I come back from overseas. That is if I am still around to come back.

Well take it easy and write when every you can. Kiss my sister in law for me and take care of yourself.

Love Bud

P.S. Tell Mom I got her letter today and ask her if see had that money insured, because I never did get it yet.

Chapter 5
TRANSFER TO CAMP MACKALL, NORTH CAROLINA

Bud transferred to the Communications Section, Regimental Headquarters. It appeared it would be a great place to be with his new Section Sergeant, Master Sergeant Dubois

Recorded in Paratrooper Odyssey (1985), Camp Mackall was not much different from Camp Toccoa, but bigger, and on level ground. There was no Currahee Mountain. Everyone was quartered in the same one-story non-insulated hutments, heated with coal stoves. Passes to town were allowed, but after a few trips to the local small towns most troopers decided they might just as well stay in camp. The newly established Airborne Center at Camp Mackall was now prescribing training doctrine for the entire Army. One of the new requirements for parachute units was a 300 mile flight, followed by a night jump, and a three day ground exercise. Bud describes the 300 mile flight, jump and field exercise in his letters. These operations took place in January and the weather turned cold, with sleet, snow and freezing rain. Men became walking icicles and found that the standard jump suit, and boots were not very practical for cold weather operations.

Letter to Mom from Harland "Bud" Curtis

Camp Mackall, North Carolina
Regimental Headquarters
Postmarked August 11, 1943

Dear Mom,

Well here I am at Camp Mackall; we got in last night at 8:30 P.M. I am sitting in a radio class now. They are talking about something I ready know so I thought I would write and let you know I'm here. We came up here in trucks. It is about 275 miles from Camp Toccoa.

My address will be the same except you put Camp Mackall, North Carolina. Dean Hildreath lives about one mile from me. This is a pretty big camp. They have about 20,000 or more fellows here. They have all the glider troops and a lot of paratroopers here. I won't jump here. I will go to Fort Benning because they have much better training equipment. As far as I know now, I will begin jump training by the first of September.

I went out to find Dean this morning. Boy it is like hunting a needle in a hay stack. I finally found his barracks, but he was out on a 18 mile march. I left a note on his bed and told him where he could find me.

The barracks are better, but they are a lot bigger. There are 48 fellows to one barracks. I didn't send my radio home. I have it here with me. I will hang on to it until I go somewhere I can't take it with me. I sure hope you had that money insured, because I never did see it. I got your telegram. Did you send the money by letter or wire?

They had my name marked that I was at Fort Benning by some mistake and they sent some of my mail there, but I am getting it sent back. Maybe it went there?

It is pretty hot here and there is a lot of sand all over just like a beach. They have a lot of PX's (Post Exchange is a store that sells everything a soldier would need) and shows scattered all over, and they're pretty nice. We don't have any mountains or hills to run up around here any way. I hope pay day hurries around because I am going to go to Norfolk, Virginia, and then I am going to go to Washington D.C. just to see the place.

If you ever find time you can send me a dog tag chain; I broke this one. Did you ever find out Roger's address (Roger Hormer a neighborhood friend) Where is Afton at? And where is Bob Douglas? Has Bill joined the Navy? Let me know how everyone is.

I will write as soon as I hear from you. See you then.

Love, Bud

Letter to Folks from Harland "Bud" Curtis

Camp Mackall, North Carolina
Regimental Headquarters
Postmarked August 11, 1943

Hello Folks,

I got Dad's letter and telegram today; thanks a lot,. I've been broke for a few days now. I owe most of this, but if I am careful and don't go anywhere I can make it last until payday. It really came at the right time.

I saw Dean the other night. He is about 2 miles from me. He is still the same, only he looks a lot better with boots and wings (Dean Hildreath served with the 101st Airborne Division and made combat jumps on D-Day, June 6, 1944 and in Market Garden jump in Holland).

This is really a big camp. Air planes fly all over the place and you can watch the troopers jump. It really looks good to see all those parachutes open up. There is about 2 or 3 regiments of glider troops here too. The nearest town is about 20 miles away. I could get a pass tonight but it is raining pretty hard. I don't know where the heck we are at here. I think this camp was set down in the middle of some forest or something.

The airborne insignia at the top of the page is what we wear on our sleeve. It is really pretty. I don't think much of North Carolina from what I've seen of it. I'll be glad when I go to Fort Benning so I can go to Atlanta again. It won't be long now. Somewhere along the first part of September. There is a fellow in our barracks who is going to jump this Sunday. He is a qualified jumper, but they have to jump every so often to keep their jump pay. He doesn't say anything, but you can see he is sure sweating it out. I can't blame him. You can't tell what will happen after you leave that door. It only takes 5 to 8 seconds to come down if the chute doesn't open. I imagine you can think a lot of things over in 5 seconds though I'll sure be glad when I can write home and tell you all about it from my own experience. A million guys can tell you what it is like, but you'll really never know until you do it yourself.

There is a lot to do in this camp so it doesn't make much difference if I get out or not. I think it would take a weekend pass just to find your way out. Thanks again for everything. Write when you can

Love, bud

Letter to Folks from Harland "Bud" Curtis

Camp Mackall, North Carolina
Regimental Headquarters
August 15, 1943

Dear Folks,

Here I am in Norfolk, Virginia. Boy am I tired I haven't had any sleep for two nights and when I get back to camp it will be just about time to fall out. I'll make it though. When I stop going, the sun will come up backwards.

I saw Pat. It sure surprised her. I am in the U.S.O. (United Service Organization) lounge at the station, I am waiting for my train. They have free milk and magazines to read. A place to write. These U.S.O's are really alright.

Well I have seen the Chesapeake Bay, and the Atlantic ocean now. How about that. I sure get around don't I. It didn't cost much at all to come here. I sure like the travel. I guess you know that by now though.

Norfolk is full of sailors; I am the only paratrooper in the whole city. I'm a one man circus with these boots. People turn around on all corners with a puzzled look on their faces wondering what I am. Well enough for now.

Love as always,
Bud

Letter to Mom from Harland "Bud" Curtis

Camp Mackall, North Carolina
Regimental Headquarters
Postmarked August 23, 1943

Dear Mom,

I was beginning to wonder if you had got my letter with my address because I haven't heard from you for quite a while until I got your letter yesterday. Thanks for the $10.00 I can use it right now. I will pay you back as much as I can with this months pay, but I can do much better soon as I start getting that jump pay, yes I got the other money order and Dad's letter so don't worry about it. I really needed it, I had borrowed quite a bit from the fellows.

Well I guess I will be at Fort Benning right soon now; about the first of September if things go right. Then when I make all my jumps I will either get a 15 day furlough then as I might have to go to radio school for 6 weeks there and then get the furlough; I don't know yet. This 15 days includes traveling time and everything. The first Battalion is all on furlough now. You might see some of the fellows there in town. There is quite a few fellows from California in this outfit.

I haven't been doing much to speak of. I am still going to radio school and so I get it pretty easy just setting around while the other fellows in Company "F" are being worked to death. They go out on long marches and a lot of night problems. I sure feel sorry for them, but if I had of stayed with them I would have a rating and would be getting another soon, but I would rather be here learning something.

I sure am glad I got in that radio class at Poly, because I am in the top of the class here in code and it helped me get the radio section instead of some other branch of communications.

I got your other letter today. I guess I will get the cookies pretty soon. I sure hope so. Time sure does go slow. I will be glad when I get to Fort Benning and get my wings and a furlough; then I don't care what they do with me then, because the next thing will probably be overseas duty, maybe maneuvers first.

Well I really haven't much to say so I'll let it go for now. Thanks for everything and I'll be seeing you pretty soon now. Love, Bud

Letter to Mom from Harland "Bud" Curtis

Camp Mackall, North Carolina
Regimental Headquarters
Postmarked August 29, 1943

Dear Mom,

Here I am again. I just got off K.P. (kitchen police) over Saturday and Sunday. It is Friday now. Oh well the milk was good and I was hungry so that's that. I really don't have the money to go out this weekend anyway. I guess I will be working pretty hard over this weekend, but I'll just have to hold out for one more week of this place I will be at Fort Benning starting jump training. The latest rumor is that we won't have to take 6 weeks or radio school until after we come back on furlough. So I think I will be home on Bert's birthday. Boy there is a big electrical storm going on outside and the lights keep going off and on. Mostly off.

Oh be sure and call up the Press and tell them to send the paper to Regimental Headquarters. Not Regimental hospital. I got the whole month of August's papers today and I guess it will take me a month to read them all.

I'll save all the money I can to come home; maybe I can borrow some from the Red Cross. If not I will just spend my time on the east coast. I don't want you to send me any more. You can't afford it and it is hard for me to pay it back. I'll let you know more about that later.

I don't know if Dean is still here or not. I haven't seen him all week and he is going off to Tennessee next week for maneuvers as far as I know.

I'll sure be glad to get to Fort Benning. It will be plenty rough, but it will be something new and different and I sure get tired of staying in one place. I wish they would ship us across now and get thing over with. I'll let you know how I turn out on my jumps and everything.

You should hear the thunder and lighting and the rain is coming down in buckets. These darn lights aren't working so good so I am going to let this letter go for now. I am sorry I haven't much to say anyway so write soon and let me know what is going on in California.

Love, Bud

P.S. I sure like your cookies they really tasted good, but they didn't last long with all these hungry guys around. I get my share of theirs too though.

Letter to Mom from Harland "Bud" Curtis

Camp Mackall, North Carolina
Regimental Headquarters
Postmarked September 3, 1943

Dear Mom,

Here it is Thursday (Maybe August 26, or September 2, 1943) and we are getting ready to leave for Fort Benning tomorrow. By the time you get this letter I will be at Fort Benning starting "B" stage which consists of a lot of technical things, tumbling, mock up towers, calisthenics, etc. Then "C" stage which is the next week is packing parachutes, high towers, free jump from 250 feet off the high tower with the chute already opened, so you can get the idea how it is and can practice your tumbling. Then they have another thing they hook on your back and carry you up on a cable like at the fair to 250 feet. You are just laying out face down to the ground. Then they say pull you cord. You pull it with your right hand and change it to your left and at the same time you count 1000, 2000, 3,000. You fall about

20 feet and then you are stopped with a nice quick jerk. The fellow says that thing is worse than going out of the plane. Two lieutenants were killed on it recently and they may not use it anymore. I hope they do though. It sounds fun.

Then "D" stage is when I jump from the plane. Oh, boy. That will be the hardest door I will probably ever go through in my life. Then we come back here and leave on furlough.

It will be about the first of October when I get back here on a few days sooner. I want you to send my suitcase here. But I will write later and tell you when to send it because I have to figure out the time so that I will be here and they won't be sending it to Fort Benning while I'm there.

My address will be the same as now, but leave off the A.P.O. 452 and address mail to Fort Benning, Georgia. I will be there Sunday (Maybe August 29 or September 5, 1943) so any time after that, that's where you will address my mail too.

After paying back all the money I owed and getting a few clothes altered so they will look nice when I get home I have $20.00. So until I find our for sure if I can barrow some form the Red Cross, don't plan too much on me coming home, because I don't think I will get paid on time next month because they have to figure out the jump pay or something. Some guys have gone 2 to 4 months without pay after they have jumped. It is nice too get that money all at once, but there is a lot of things I could use it for in the meantime. Oh another thing just as soon as I start "B" stage I start earning that $100.00 a month.

You remember me saying I would be on K.P. (Kitchen Police) Saturday and Sunday. Well I got my Master Sergeant to get me off of it and then he invited me to a party he was going to in Charlotte, N.C. so the Sarge and me took off to Charlotte which is about 100 miles from here. I had a good time and everything was going swell until I walked out on the back porch. It was really dark, I couldn't see a thing. Well the steps turned left and I kept going straight and walked off the balcony which was 15 feet to the ground. I turned a couple of loops, counted 1,000, 2,000, 3,000, but remembered I didn't have my parachute. I hit the ground and shook it like an earthquake. What an awful dent I put in the ground. It didn't hurt me much, being as I lit on my head, so I got up and after a big argument everybody decided I was still alive although I didn't look it. It was a wonder I didn't kill myself, but I guess that is impossible. I really think I should get my wings now for making a might jump without a parachute. I had a swell time and the Sarge is the head of the communications and all and I'm in pretty good with him now which will help out a lot.

I have to pack my clothes and things I'm taking, so I'll stop now. Write soon, and I'll let you know how I make out.

Love, Bud

P.S. You had better stop the Press Telegram because I will be moving around a lot.

In speaking with Bud on Wednesday, January 12, 2005, he stated that he and Master Sergeant Dubois went to Charlotte, N.C. for a weekend pass and were at some house that was off limits. The reason he walked off the balcony was because he was running from the M.P.'s who burst into the house and were after the soldiers. Bud ran out the back door. The steps turned left and he kept going straight and fell over the balcony. Today Bud says he landed on his back not his head as he said in his letter to his mother. He was bruised but okay. The M.P.'s took him and Master Sergeant Dubois into custody. When they got to the M.P. station in Charlotte, Bud was being very polite and not antagonizing the M.P.'s. Master Sergeant Dubois on the other hand was giving the M.P.'s a bad time and trying to exercise his authority as a Master Sergeant. The M.P.'s were not buying it. They put him in the stockade and let Bud go back to camp with no charges because he was polite.

Chapter 6

FINALLY, JUMP SCHOOL

Fort Benning, Georgia

In late summer of 1943, Regimental Headquarters moved out from Camp Toccoa to Fort Benning, Georgia for parachute training. The parachute school was just down the road south of Atlanta located in Columbus, Georgia. A six hour truck ride from Camp Toccoa. Fort Benning was and still is a model for military efficiency in the highest sense of the term. Thirty-five foot mock airplane fuselage towers doted the camp that was similar to the ones at Camp Toccoa. Landing fall trainers, the suspended harness, the plumbers nightmare, and other instruments of torture were there to train troopers how to jump correctly. In the main post area were four 250 foot steel parachute towers, modeled after one that had been at the New York World's fair in 1940. These towers still stand today and are still used to train paratroopers. Each parachute school instructor was a rugged physical specimen with an air of authority and command that was envied by most officers. In September 2003, Bud returned to Fort Benning for the first time since 1943. As he walked the field where the jump towers were he related, "In 1943 I was on this field in front of the 250 foot towers when one of the jump instructors told me to fall down on the ground and crawl. I followed his instructions really not knowing why." Then he said, "Your leg has just been blown off, now craw with one leg, now your other leg is gone crawl with no legs, now you have lost one arm, now crawl." When I asked Bud if he ever thought he would be back on that field sixty-one years later his reply was, "Never." Bud and the other troopers were taught the "Airborne Creed." This they learned and lived by. Below is the "Airborne Creed" which is:

The Airborne Creed

I am an Airborne Trooper! A paratrooper!

I jump by parachute from any plane in flight. I volunteered to do it, knowing well the hazards of my choice.

I serve in a mighty Airborne Force - famed for deeds in war - renowned for readiness in peace. It is my pledge to uphold its honor and prestige in all I am - In all I do.

I am an elite trooper - a sky trooper - a shock trooper - a spearhead trooper, I blaze the way to far-flung goals - behind, before, above the foe's front line.

I know that I may have to fight without support for days on end. Therefore, I keep mind and body always fit to do my part in any airborne task. I am self-reliant and unafraid. I shoot true, and march fast and far. I fight hard and excel in every art and artifice in war.

I never fail a fellow trooper. I cherish as a sacred trust the lives of men with whom I serve. Leaders have my fullest loyalty, and those I lead never find me lacking.

I have pride in the Airborne!
I never let it down!

In peace, I do not shirk the dullest duty nor protest the toughest training. My weapons and equipment are always combat ready. I am neat of dress - military in courtesy - proper in conduct and behavior.

In battle, I fear no foe's ability, nor underestimate his prowess, power and guile. I fight him with all my might and skill - ever alert to evade capture or escape a trap. I never surrender, though I be the last.

My goal in peace or war is to succeed in any mission of the day - or die, if needs be, in the try.

I belong to a proud and glorious team - the Airborne, the Army, my Country.

I am its chosen pride to fight where others may not go -

to serve them well until the final victory.

I am a trooper of the sky! I am my Nation's best! In peace and war I never fail.

Anywhere, anytime, in anything - I am AIRBORNE!

The parachute course consisted of A, B, C, and D stages, each a week long. Physical conditioning was emphasized. The 250 foot towers were used in C stage and D stage consisted of 5 parachute jumps. Bud explains these stages in his letters to his mother. Bud and the others who had spent so much time at Camp Toccoa preparing for this parachute training were well prepared. The months of pushups, speed marches, calisthenics, and

Chapter 6 - Finally, Jump School

running up and down Currahee all paid off for the 517th troopers. These men were more prepared for jump school than any other unit. The school instructors, accustomed to classes of men who had only finished a regular basic training were astonished to find that the men of the 517th were in just as good a shape as themselves. Extra miles of running and pushups were piled on the men of the 517th, but their extra requirements had no affect on these men. They were ready, hardened, and as tough as their instructors. It was finally conceded by the instructors that there was no way they were going to get the 517th to cry "Uncle."

Bud Curtis revisits Fort Benning, GA, September 2003 the same 250 foot jump towers he jumped from in September 1943 are still being used today by the Army to train paratrooper

Parachute training at Fort Benning, Georgia in 1943, pictures courtesy of Grant Dougherty

FIRST JUMP

Letter to Mom from Harland "Bud" Curtis

Fort Benning, Georgia
September 2, 1943

Dear Mom,

It sure was good to hear you and Dad over the phone Sunday. I went to church this morning. It sure was cold in Atlanta; I didn't know it could get that way after all the heat there was when I was in Toccoa.

Well I came out swell on my first jump Monday. Gee what a thrill we put on our parachutes, then went to the inspection line and into the plane (a C-47 Transport). Well we taxied out on the field and up into the air and circled the field a couple of times. I was the 11th man in the first stick. I can honestly say I wasn't a bit scared, but there were some awfully white faces in that plane. When we got to 1200 feet the jump master says "Get Ready." We take our anchor line snap fastener in our hand. "Stand Up." We stand up grabbing the cable that runs the full length of the ship. "Hook Up." We hook our snap fastener onto the cable. "Check Equipment." We look at each other's pack to make sure it is okay. "Sound Off Equipment Check." 12, okay, 11, okay, 10, okay, 9, okay, 8, okay, 7, okay, 6, okay, 5, okay, 4, okay, 3, okay, 2, okay, 1, okay. "Stand in the door." He shouts. We close up and he says, "Are You Ready." YEAH, we sound off. "Well Lets Go." Then is when you get a funny feeling seeing those guys in front of you disappear. Here I come shuffling up to the door; plant my right foot down hard, pivot in the door, jump out about 6 feet, make a half left body turn, ducking my head so the connector links wont knock you in the head. I count; 1,000, 2,000, then my feet are sticking out in front of my face and it feels like the air is churning around inside your head and all of this happens between (counting) 2,000 and 3,000. Wham, the chute opens and I get an opening shock that liked to shook my teeth out. Then it seems like you are just suspended in air for a while and you are oscillating back and forth. It's quiet and peaceful up there and it sure looks pretty, all those other chutes below me.

Then a big grin comes over my face and I think, I'll be darned if I thought that thing would open. There I am hundreds of feet above the ground with not a care in the world. It is quiet like a graveyard and I can hear the voices of the instructors running through my mind. Check oscillation; make a body turn so the ground is coming toward you; keep your feet together; knees slightly bent; prepare to land. LAND. And at the last 100 feet the ground is coming up to meet me fast, but I land like a feather and make a right front tumble; spill the wind from my chute and take my harness off and head off for the truck to take me back.

Then my head begins to clear up and I think what I did and I feel proud of myself for having the nerve to go out of that door and I am glad I had the guts to stick out all of this training I've had to go through because it is really worth it. That's my kind of work. You will never realize what a thrill it is.

Some of the fellows didn't do so good. Five broke their legs, and a few sprained ankles. A couple of fellows got tangled up together and were coming down fast. They pulled their reserves and came in okay. I don't think we will jump today, because it has been raining and we will make two jumps Wednesday.

It still doesn't seem like I did jump; it's more like a dream because everything happened so fast. It won't be long until I will be home now if I come out as good on the next four jumps as I did the first, so keep your fingers crossed until Saturday.

Tell everyone hello. I have to fall out in about 10 minutes so I had better get this mailed.

See you soon, Love Bud

Letter to Mom from Harland "Bud" Curtis

Fort Benning, Georgia
Regimental Headquarters
Postmarked September 9, 1943

Dear Mom,

Here I am in Georgia Again. This place is big as a city; I think they took about half of Georgia and put a fence around it and called it Fort Benning. They have everything here, even WAACS. We are living in tents; they call it the frying pan area, but I believe it is the fire.

I am half way through "B" stage now and we have been learning how to pack parachutes in the morning and in the afternoon we have mock towers, tumbling, and a lot of other contraptions. Next week we will start "C" stage and that is all the 250 feet high towers. They have a parachute already opened attached to a frame which is hurled up to 250 feet and when you get to the top they release you and down you come. We have to do this at night time too. More fun.

After "C" stage is "D" and then we jump. Boy that is what is going to be a real thrill. I guess you can imagine. I bet the women would like to get a hold of one of these parachutes; there is sure a lot of silk in one of them. These sergeants here sure keep you on the ball; they are all a bunch of supermen, and what ever they say goes. I was watching a school of officers in "A" stage. They were doing push ups and these sergeants walk around yelling at them. Captains, Majors, Colonels, even if a 4 star General goes through this school the sergeant is the boss. After the things I've had to take from some of these officer it sure sounds nice to hear one of the sergeants bawling out a Major. "What is the matter are you tire, can't you take it", and there all so tired they can't see straight.

Well if everything goes right I will be a qualified paratrooper soon and it won't be long until I come home if I can raise the money somewhere.

Love, Bud

P.S. Send my suitcase so it will be at Mackall about the 25th or before the first of October

Bud and others learned to pack their own parachutes.
Then they jumped with those self packed chutes

Letter to Folks from Harland "Bud" Curtis

Fort Benning, Georgia
Regimental Headquarters
Postmarked September 14, 1943

Dear Folks,

I got your letters yesterday; it took them a little time to get organized, but now I am getting the mail regularly. I started the first day of "C" stage today. I was hauled up on a 250 feet tower and was released; boy, what a thrill to come floating down. This whole place is just like an amusement zone. People would pay money to ride on that contraption they have here. That is if they had nerve to do it.

A class went through "D" stage today, and a whole regiment was making there first jumps. Planes were flying over the field all day dropping paratroopers. It sure looks good to see all those chutes open up. On one stick they jumped today there was a little parachute above all the rest and it was a darn dog. I'll bet he was wondering what it was all about. I guess if a dog can do it I won't have much trouble.

Well a week from today and I will be up there. I can hardly wait. It only takes about 8 seconds to come down if that chute doesn't open so you have to think plenty fast and get that reserve into action and about 3 seconds will be gone before you know you have to. It is not a pleasant subject, but it makes you think what can happen. Last week a colonel was killed. His main chute got tangled up and he must have passed out or something because he never got to his reserve. Boy, some of the things you hear around here. It is no wonder fellows sweat out his jumps, but if you do what the instructors teach you there is no possible way that anything can go wrong. If I had paid as much attention in school as I have here I would be a second Einstein. Don't worry about me, because I'm not worried. I think it is a chance of a lifetime.

I don't know if I made you think all this is dangerous, because it is really not. The only guys who ever get hurt are the ones that don't do what they have been told so be expecting me home soon now, because it won't be very long until I can dig into that refrigerator again. Fill it up good, okay.

I'll call you up next weekend about the same time as I have before. I'll write a long letter soon. Until then,

Love Bud

Four Day jumps and one night jump to finally qualify as a Paratrooper

As World War II approached the Army became aware of the importance of being able to deploy combat infantry men from the air to the ground quickly. Parachuting was determined to be the best way to accomplish this mission. The Army knew this new form of combat deployment would take men who were extremely brave and risk takers. In 1940, the Army began testing employing soldiers from the skies. They found their strategies were sound and this form of striking the enemy from above proved to be an effective way to gain superiority in battle.

Once parachute training was approved by the Army they needed a way to identify members of this unique organization. It needed to be a badge that signified the art of military parachuting. Then Captain William P. Yarborough of the 501st Parachute Battalion designed a Ojibway Thunderbird on a silver shield with a motto of "Geronimo." He had hoped this design would be adopted as the new parachutist badge and would be on every paratrooper's uniform. Higher brass wanted something more distinctive. At the time Army aviators had a shield with wings extending from it to distinguish them as pilots. Everyone recognized this insignia of a pilot. No way did senior infantry officers want their paratrooper badge to resemble pilot wings of the Air Corps.

On March 1, 1941 then Major (MAJ) William Miley was commanding the 501st Parachute Battalion. He called in Captain (CPT) William P. Yarborough and ordered him to Washington D.C. to report to the Adjutant General Office with temporary assignment in the Office of the Chief of Infantry. MAJ Miley told CPT Yarborough to not come back to Fort Benning, Georgia until he had an Army approved parachute qualification badge. At the time many different badge designs had been supplied by the G-1 Department of Heraldry in Washington, D.C., but each one was promptly rejected. One designed showed a deployed parachute around folded wings which almost gave it a funeral type attitude. CPT Yarborough had his work cut out for him as he arrived in Washington, D.C. Not only did he have to come up with a parachute design that would represent all paratroopers in the Army he had to make sure it fit the parameters that were given him by the bureaucrats that were not Airborne qualified.

After fifty (50) different tries of submitting badges to the Department of Heraldry, CPT Yarborough finally came up with a design that he thought could be accepted. CPT Yarborough drew a set of wing tips that were supporting a parachute canopy symbolizing powered flight from which always preceded the paradrop. This design had promise and maybe it just might win the approval of the Army brass. CPT Yarborough described, " I walked the approved design from the Department of Heraldry in and out of every office in the War Department who had to sign off on its design. I would wait doggedly until each senior action officer got it in their in basket. As soon as they approved it, I would walk it down to the next person, then the next until I had every signature approving the parachutist badge design." The parachutist badge was formally approved on March 10, 1941.

Finally, airborne troops had a parachutist badge everyone could be very proud of. CPT Yarborough then went to The Bailey Banks and Biddle Company of Philadelphia, PA. They he had them make the first 350 Sterling Silver wings. CPT Yarborough stated, "I camped out on their doorstep until I was able to walk away with those 350 sterling wings. These I carried triumphantly back to now promoted Lieutenant Colonel Miley at

Fort Benning. All of these first wings bear the mark of BB&B on the back, and they are collector's items today."

CPT Yarborough felt the silver wings needed a little color and perhaps they were a little on the small side. CPT Yarborough then said, "I designed the first felt backgrounds for the wings. For infantry I designed a background of blue and red for artillery with silver piping around it." CPT Yarborough continued, "Later I took a patent out on these wings in order to protect the design from wrongful exploitation, and to keep the quality high. Patent #134963 was granted on February 2, 1943 for a period of three and half years. I never obtained a single penny from sales of the wings, nor from any commercial use. This was not my objective." William P. Yarborough became a great fighting soldier of the United States Army and attained the rank of Lieutenant General. He is solely responsible for the design and development of the original paratrooper wings, paratrooper jump boots, and paratrooper uniform.

To every man who joined the paratroopers, these wings became his most coveted possession. Each of these newly qualified paratroopers had to work extremely hard to receive this prestigious award. By having jump wings on your chest you proclaimed to the world that you were someone special. Someone who had gone beyond the normal call of duty, you had volunteered to jump from a perfectly good aircraft into a combat situation where you could be killed before you every landed on the ground. Soldiers everywhere saw the wings, and jump boots of these paratroopers. They looked in awe and know that this person was somebody they did not want to mess with. They were and are the toughest of the tough.

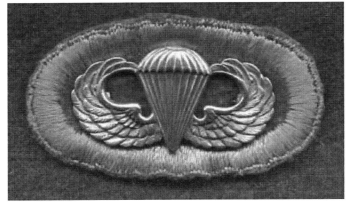

This is the infantry oval that all 517th paratroopers wore behind the wings on their uniforms

Chapter 6 - Finally, Jump School

Chapter 7

Return to Camp Mackall

Letter to Mom from Harland "Bud" Curtis

Camp Mackall, North Carolina
Regimental Headquarters
Postmarked Thursday, October 7, 1943

Dear Mom,

I just got your letter and I am going to answer it now as I am going out on a field problem tonight and won't be back until Saturday (Maybe October 9, 1943). We were going to make a night jump, but the plans have been changed and they are going to take us out, where we would land if we jumped, but only it will be in trucks. It is a little safer but not as exciting. This is for the communications only. We are going to capture our airfield and set up communications and everything as if in a real battle. We start at 12 midnight, and I won't get any sleep tonight or tomorrow, and maybe not tomorrow night. We are going to live off K and C rations only, and if the weather stays as cold as it has been I am afraid it will be a shade rough. Then we walk back to Camp Saturday which will be about 20 miles.

Charles Daigh lives about 100 feet away from me if he is in the First Battalion. So he won't have any trouble finding me (Charles Daigh was from Long Beach. Bud never knew him growing up in Long Beach. Bob Steele introduced Bud to Charles in the 517th. Bud said Charles liked to brawl and was in fights all the time. He was a tough guy). *I hope you sent that suit case by now as it won't be long until I will be needing it. Probably in the next two weeks. Do you think Bert will let me use his car? I haven't driven a car since I left there; I guess I still know how. How is the gas problem now? Does he have any coupons? I'll get some someplace if he hasn't got any. Tell Dad I got the money, and thanks a lot.*

I put on the first pair of long underwear yesterday, that I haven't worn in a long time and they are sure needed here now. It would be nice if I could fly home, but I am afraid that I might that I might get the reservation cancelled and then I would lose out on a lot of time. Maybe I can do it coming back.

Write soon and I will probably get your letter when I get back from this problem. I will call you up Sunday if I am some place where I can. If I don't call up before 8:00 A.M. Sunday morning don't wait around and ruin your day as I won't call after then.

See you soon I hope

Love, Bud

P.S. *Thanks for the stamps, I was out of them.*

Bud at Camp MacKall, center photo wearing newly earned Jump Wings and Good Conduct ribbon

This letter tells about the long hours of training, nowhere to go to get away from the orders of the sergeants and the life of a paratrooper training for war. Bud has been in constant training, far away from home for six long months of his life. The Army keeps telling the soldiers they will get a chance to go home, but as of yet the opportunity has not arrived. In these next few letters you will read some of the discouragement Bud is experiencing being so far from home, yet at the end of his letters he acknowledges his great pride of being one of the elite of the Army, a paratrooper, the toughest of the tough. Bud talks of the grueling experiences of the obstacle course with machine guns firing, and explosions going off while crawling through dirt and mud.

Letter to Mom from Harland "Bud" Curtis

Camp Mackall, North Carolina
Regimental Headquarters
Postmarked October 3, 1943, Letter says Saturday, maybe October 2, 1943

Dear Mom,

Here I am in Camp Mackall again, we got here last Wednesday night and I haven't been anywhere I could call up. They told me I had a telephone call the other day; I guess it must have been you. I don't know where I was at; probably at the P.X. (post exchange).

Well the way they are working the furlough deal is they have about 55 fellows in this company that are eligible for furlough. So they mark numbers on slips of papers and then we drew them out of a helmet. My number is 22. They let 27 out at a time and right now there are 22 out so from the drawing today they let the just five go. Now as the fellows come in they let that many more go out so it will stay an even 27 men out at a time. It will probably be a couple of weeks or maybe more before I will get my furlough. I tried to buy a lower number, but everyone seems to want to go pretty bad.

I'm getting pretty well fed up now with everything. It's my turn for K.P. (kitchen police) tomorrow so that means I can't even have a weekend pass. I only came in this outfit to get my wings and now that I have got them I don't care if I never see the paratroops again.

They are talking about a 40 mile hike some place soon so I guess I am going to get hooked into that. A guy can take just so much and I have been taking it twice as long as most the guys here as I was one of the first ones in. If it was just me that felt this way I would think there was something wrong, but everyone else is feeling the same way so it is not just me.

Well, enough for now I am just getting myself into a mean mood or something and that just makes it worse. After all I haven't anybody but myself to blame for being here as I had to volunteer for this and no one made me. I guess I might as well be satisfied because I am proud to be in the toughest outfit that has ever been organized, but I have taken everything they had and I am satisfied with the fact that there isn't anything that I can't do.

It must be that something inside of me that makes me want to go places and do things different that made me want to come into the paratroops. Now that little something is making me want to take off some place else' another outfit or something. I would like to be able to get into the Navy for a while.

I sure went over a couple of rough courses yesterday. In the morning we went out 4 miles to a infiltration course. I had to crawl 100 yards through dirt and sand while machine gun bullets were whistling all around me. Then I would crawl between a couple of holes in the ground and they had dynamite in the holes which they would blow up when I got between them. It was a shade rough and I was spitting out dirt all day. But that wasn't nothing compared to the next course. It was a mile long and I had to run as fast as I could all the way holding a rifle and boy that rifle gets heavy toward the end. I would run a long this path and someone would take a shot at me and a bullet would hit about a half foot from me and I have to hit the ground fast and fire where he is at. Then I would get up and tear along and hit a trip wire and a booby trap would blow up beside me and then I pick myself up and take out again and I come to a hole full of mud about 20 feet across and so stinking dirty a pig would turn sick and walk off. I jump in and I'm up to my neck in this slime. Your so tired now it doesn't make any difference. Then someone takes a shot at you from the bushes and there is a house where the bullet is suppose to have come from, so I take a hand grenade and charge the house and throw it through the window and hit the ground. Well all along it is like this machine gunners shooting around your ears, guys ambushing at you; and dummies you have to charge with bayonet, and these stinking mud holes a few barbed wires and wall to go over and all the time you are tripping over wires and having bombs blowing up around you.

Well I just got off K.P. It cost me $5.00, but to get out and go somewhere else is worth a million if you have it; although I have been spending too much. Next pay day I get $150.00 so I will be able to start paying you back.

Heck, the guy I had to work K.P. for me just came back and he has decided to go somewhere, so now I will be working from 6:00 A.M. until 9:00 P.M. What a way to have to spend Sunday.

Well good; now I have got another guy to do it for me so I am going to take off for somewhere. I don't care just as long as I don't see this place till 6:00 A.M. Monday morning. Well I had better get going now. Have Dad send me a key for that lock and another extension cord. It doesn't have to be as long as the other one. If I find the guy who stole the one I had I'll kill him. That is the Major's orders. If you catch anybody stealing to just bring him to the Major's office while he is still warm. So it is not such a good idea to get sticky fingers around here.

I guess this letter doesn't make much sense; I have just been jabbering about a lot of little troubles, so just forget them. Don't forget to send my suitcase right now if not sooner.

Love as always

Bud

Thanks for everything

Bud with his older brother Bert home on furlough in October 1943. Who would know when they would ever see each other again.

Letter to Mom from Harland "Bud" Curtis

Camp Mackall, North Carolina
Regimental Headquarters
Postmarked October 29, 1943, Written probably Tuesday October 26th

Dear Mom,

As far as I know I will get my furlough this Saturday, but anything can happen between now and Saturday. I would sure like to see everyone again, but I am beginning to lose all interest in whether I get a furlough or not. This Wednesday we have to take a divisional ABC test. It is a physical test of just about everything (Bud said he had to do sit ups, push ups, running, and chin ups. When Bud got on the chin up bar he began to show off. He swung around the bar backwards and forwards doing circus type tricks. He would swing backwards hanging by his knees. Then let go twist around and grab the bar with his hands and swing down. The Captain saw this and told everyone to come over. Bud had to perform it again for all of the troops.) *It will be pretty rough. This Thursday we have a communications problem. We are going to jump with our equipment and radios. I will have six jumps in after Thursday. I hope nothing happens to keep me from leaving Saturday. If nothing happens I will send a telegram Saturday telling you I am coming. If something does go wrong I won't be able to send you one so in that way you will know if anything has gone wrong. I don't particularly like this jump coming up Thursday, and having a furlough Saturday, because there is always that chance of spraining an ankle and I don't want to come home limping. It would make people around California think the paratroops are dangerous or something.*

After I come back I wouldn't care if I jump everyday, because I might break a leg and then I would get another furlough for 30 days. I'm just kidding, but it is a thought that might work to a good advantage maybe.

How is the weather in California? I guess most everyone I know is gone by now aren't they? If Bert can stay on with his job that he has he will be awful dumb to quit. He is a lot better off right where he is at.

About that telegram; it might be Sunday before I get to send it; so if you don't get it by Sunday I will not be coming home this week.

Well hope for the best.

Love

Bud

Bud probably left Camp Mackall on Saturday, October 23, 1943. He was given fifteen days furlough. He booked passage on a train and rode in a chair from Charlotte, N.C. to Los Angeles, California. It took him five days to get arrive. There his parent met him up and took him home. Bud only stayed home for five days. He probably left Long Beach around the first day of November 1943. He went to Santa Ana, California and was able to get a military hop on an airplane to El Paso, Texas. On the flight he met a soldier named "Smitty" from the Air Corps. He knew all of the tricks to flying on military aircraft. He told Bud that they would fly to El Paso then catch another flight to an Army Airfield on the Louisiana Texas border. From there they hitchhiked maybe 100 miles to some town where they caught a train to Baltimore. There Bud stayed with "Smitty" for a day and half then took a train back to Charlotte, N.C. See the letter Bud wrote Sunday, November 7, 1943. He arrived back in camp with plenty of time. Bud remembers the weather being very cold.

Bud home during his first furlough in Oct 1943 with his Dad, Bert Sr, and brother Bert, Jr. At right is Bud with his mother Luciel.

Letter to Mom from Harland "Bud" Curtis

Camp Mackall, North Carolina
Regimental Headquarters
Sunday, November 7, 1943

Dear Mom,

    Well here I am back again. I had a swell time while it lasted but now I have to get use to being a soldier again. I won't be sleeping until noon again for a long time.

    I finally got out of Santa Ana the second time at 5:30 pm Tuesday and got in El Paso at 10:30 pm and slept there for the night. It was sure a thrill being up in the nose of that B-18 flying over Phoenix City, Arizona at 10,000 feet at night time. It looked like a miniature city all lit up. I got a B-25 out of El Paso and was in De Ridder, Louisiana in the afternoon Wednesday. It was a desolate spot so Smitty and I (Smitty is a fellow who left the Santa Ana field the same time I did. He was going on furlough from there to Baltimore and we got to be pretty good friends) hitch hiked up to Shreveport, LA about 130 miles to a better airport but the M.P. wouldn't let us in until morning so we slept on the bench all night and was I stiff when I got up. We caught another B-25 out of there to Florence S.C. but we had engine trouble and had to make a forced landing in Columbia S.C. I thought for a while I was going to have to make a 7th jump.

    We got the thing fixed and got to Florence in the afternoon Thursday. I had an extra day so I went up to Baltimore with Smitty by train and Friday night we went to a night club and he really showed me a swell time. I slept at his house that night and left at 7:15 am Saturday and then called you up in Washington D.C. I went out on the street from the station and asked a lady which way it was to town and she said that it was where she was going so she drove me all around the White House and a lot of other important places. I liked Washington D.C. and I am going up there the first 3 day pass I get. Well I made it in an hour and a half before my time was up. I worked it just right only I think I could have got away with not being in until Monday reveille, but I did get in when I was suppose to and I'll have a clean service record and that will help the next time I want another furlough. I hope it won't be too long as I really enjoyed being home. Everything went just right and Bert was swell letting me use his car all the time and Lorrain is sure nice too. I hope you didn't mind me spending so much time with Jill, but I think a lot of that little scatter brain so I hope you understand and take care if her for me until I get back O.K.

    I have your **F** book and I will send it back in this letter also, as soon as I can I will get a money order and send this money home. Some guy walked off with $1500 of the pay money so I probably won't be paid until next month. That will be $250 and I'll send it home. I want to start saving money as I am sure going to need it when I get out of this Army.

I still have my watch set on California time it is easier to check up on what people are doing there.

I must have left most of my socks home as I only have a couple pair here. Have Dad get a couple of those small batteries for that small flashlight if he can get them.

I think Smitty will come around and say hello as I gave him the address and he is stationed in Santa Ana. He can tell you more about our trip back than I could write. We had a lot of fun but it is going to be a few days until I get use to this life again. I think I could get home sick right now if I let myself, but I'll talk myself out of it.

Write often

Love

Bud

Letter to Folks from Harland "Bud" Curtis

Camp Mackall, North Carolina
Postmarked November 13, 1943

Dear Folks,

I got Dad's letter and the pictures yesterday and I guess that was all of them or did you have any more? I wonder if Bert would let me use that code sending machine he has at his apartment. If he is going into the Merchant Marines he probably won't have any use for it. You see, if I had that I could build up my code speed and that is what I need. The only trouble is that I really need it right now, and Bert will probably be having to use it until he goes in. Find out about it and if he can part with it for a while. I would sure appreciate it. Send it as soon as you can and ask Dad if he can build some kind of a box like he did for my radio to put it in as I will have to mail it home before we go on maneuvers. Also send that code key and a couple of reels of that code (and directions how to put it together).

There has been rumors that we will get another furlough before we go across and they tell me the First Battalion has already started to get a second one (furlough). But don't believe it until I am actually home again as you can't plan on anything in this Army although I am pretty sure now that we will be on this side a little longer than I had planned on. I will just have to wait and see now.

Right now I am riding around in a command car with a high powered radio in it. We are operating on code so you can see I have use for that code machine Bert has. Right now I am just a junior operator; the only thing that is holding me back is I haven't had enough code.

Also ask Dad if he can send me a socket like this (Bud drew a picture of a light bulb with a plug on it). One that I can plug in and can put a light bulb in the other end.

I really haven't anything to say so good bye now.

Love, Bud

Letter to Mom and Dad from Harland "Bud" Curtis

Camp Mackall, North Carolina
Postmarked Wednesday, November 17, 1943
Letter marked Tuesday, maybe November 9. 1943

Dear Mom and Dad,

How is everything at home now that I have gone. I'll bet you all gave a sigh of relief when that plane took off with me aboard. How does it feel to get a nights sleep again without some one coming in at the wee hours of the morning. I can imagine Bert was awfully happy to get his car back too. Oh well, I sure had a swell time home even if I did about ruin the health of everybody and the neighbors while I was there. I would give anything if I could be home for Christmas, but I am afraid that I will have to make it the next Christmas.

Have you seen anything around Long Beach that I could give Jill? I don't get much chance to look around in the day time unless it is a Sunday, then all the stores are closed. I want to get something nice. Know what I mean, I don't? Also look around and see something you like for yourself and for Dad and Bert, anyone else I should send something too. I am afraid that this year I will have to do my Christmas shopping through somebody else. So as soon as I get paid I will send the money home and you do what you want to okay.

I guess whatever you do get for Jill, you had better send it to me first so I can take a look at it and then mail it back to her from here. I am afraid there is going to be trouble if these people don't start paying me around here. I haven't been paid for going on 3 months now. Oh well, it will be nice to get it all at once though.

Guess where I am right now. Well I am sitting in a big radio command car operating the radio which has the Colonel on the other end and when things get dead I get a chance to catch up on my letter writing, but don't be surprised if this letter ends up in code as I have a pair of earphones on and any minutes there might be a message coming through. The reason for this radio is that the Colonel wants to be in contact with Headquarters at all times and that is where we are parked right now. He is in another car like this one and he roams all over the camp, and out to the ranges and if anything important comes up that he should know about I send it to him over this radio. If there is anything of importance that he wants Headquarters to know about he just sends it to me and I give it to the runner who takes it into the message center across the street. Simple huh! I do enjoy it much more just sitting here than being out in the cold someplace so it is a pretty good deal. The Captain of Communications told the wire section chief that I would make a good wireman so now he is trying to get me into that section and says if I am good enough I will be a Corporal, but I am not looking for any stripes or I would had them a long time ago. I just want to get along with people and I find out it is a lot easier to do it when you are a Private. If I stay here in this radio section I will probably get a couple of stripes in time, but not as soon as most of the ratings have been taken. So if I have my way about it I will stay right here where I'm at.

My Communications Sergeant, which is my boss and another T-4 Sergeant (Technical – 4. Rank insignia was three stripes with a "T" under the stripes) went up to a place. Union, South Carolina about 150 miles away from here last weekend. This T-4 Sgt has his home there and his folks have a farm in the country. So I was out with the cows and the other farm animals having a good time with them this weekend. He has a 1941 Studebaker (this was a car) which he goes back and forth from Camp and he has a motorcycle at the farm which he took me riding all over the country on. More fun.

This T-4 Sgt I'm talking about use to be one of these daredevil stunt men before he came into the paratroops. He used to run cars through brick walls and all those kinds of things. He is a nice guy. I don't know how I get along with this Master Sergeant so easy though as he gets pretty tough when he is on duty, but he has to be and I know some of the fellows don't like him any to well, but I do and I get along with him well. You see, that is what I mean by having a few stripes on your arm and all of a sudden people don't like you because you have to tell them what to do. No, I am perfectly happy to be Private Curtis.

Well it is about time to close up this radio net so good bye now.

Love, Bud

P.S. Has Bert heard anything more about the M.M. (Merchant Marines) and what did he say about his code machine? Tell Jill hello.

Letter to Folks from Harland "Bud" Curtis

Camp Mackall, North Carolina, Postmarked November 24, 1943
Written in left top corner, Tuesday Night, maybe Nov 16 or 23, 1943

Dear Folks,

Well here I am back in the woods of North Carolina, and playing soldier again. Gosh, but I had a swell time, but I came in about 5 hours late as I'll explain later.

I hitched hiked up to Washington, D.C. and then took a train up to New York City and got in about 9:00 A.M. Boy oh boy, what a place. Honest a soldier could live there forever and it wouldn't cost him a cent. It is sure a lot different than the South! Well first I got something to eat and then sent a telegram home to you as I had the exact sum of 75 cents. Then I went to a U.S.O. on Park Avenue to get tickets to these classy places and I met another paratrooper there and he was on a three day pass and he lived here in Manhattan. Well from then on he was my personal guide and he got me on a subway under the ground (just like a gopher) and we went out to his place and I met his mother and sister. They were very nice and invited me to stay there for the night. Jimmy and I went back to town and after going around to a few places of interest we went to a stage play which the U.S.O. had given us tickets to. It would have cost us $5.00 a piece if we had to pay for it. It was really a super comedy and I sure enjoyed it. When we got out we went to the "Stage Door Canteen". It is not very big but it is pretty nice. There are sure a lot of things to do in N.Y. and for the short time I was there I certainly got around to a lot of them. Jim called up his girl friend there and told me she was getting a girl for me. Just to be sociable, I agreed. I was expecting to get a blind date like a person usually gets. Know what I mean? But when I got over to Jim's girl's place she introduced me to this girl. Irene Sarse' (pronounced Sorsay). She was French with big beautiful brown eyes, long black hair that came to her shoulders and was about Jill's size. She was sure cute and looked like something from Esquire Magazine. I think Jill is about ready to shoot me as I told her in a letter I just wrote that I took her out. We went up to the top of the Empire State Building and you would have thought it was a hurricane the wind was blowing so hard. Boy, but that is a tall building. It is 45 feet higher than I have jumped from. What a view. Then we went to the Rockefeller Center and Radio City and a lot more places and then ended up at the theater in Radio City. It is the biggest theater in the world. All the time I was in N.Y. my mouth was wide open from amazement from so many magnificent places. I guess people thought I just came off the farm! They had a swell stage show with the Rockettes that are so famous. I guess you have read about them. It was quite late when we got out so we just went to a couple of night clubs and then called it a day (and night). I am sure glad I met Jimmy as he really went out of his way to see that I had a good time. I don't know how I met such swell people, but somehow I always do it seems like. I slept at Jim's house and got up and caught a 9:30 A.M. train in the morning. I had plenty of time to get back to camp, but I went to sleep and woke up 110 miles past where I was suppose to get off. I was in Columbia, S.C. and by the time I got back I was 5 hours AWOL (absent without leave), but I sent a telegram ahead and I had a pretty good excuse so the C.O. (commanding officer) didn't do much except say I will have to walk two hours each night for every hour I was late (with full field pack) so as soon as I make up 10 hours (meaning Bud walked a penalty tour of 2 hours for every hour he was late) *that is all and this doesn't go on my service record so I got off pretty easy I guess.*

I got your cookies and it will take the whole barracks a month to eat them (if the mice don't get them first) there're really swell. Thanks. I got the bananas, but they were very ripe. We ate most of them, but I'm afraid you had not send anymore. Darn it. I sure like bananas, but they sure take a beating in the mail. Pardon me for writing so small, but his is the last page of stationary I've got and everybody is asleep. Oh well, I probably got more on here than I could on 4 pages the other way. Thanks again for sending me the money in N.Y. and thanks for the cookies and everything. I will write again soon. Tell Dad thanks for everything. Tell Jill I think more of her than a million girls with brown eyes. I imagine she is about ready to kill me. Have you seen anything I could give her as yet? Let me know as it is getting along that time now.

Love from the AWOL kid, Bud, P.S. How did those pictures turn out? I hope not too bad!

Letter to Mom from Harland "Bud" Curtis

Camp Mackall, North Carolina, Postmarked November 25, 1943
Written in left top corner, Wednesday Night, maybe Nov 15 or 20, 1943

Dear Mom,

I got your letter today and you said you wanted me to write soon and tell you what I think of which house to get and etc. Well Mom, that is up to you and Dad as to which would be the best, so I have lived at so many places in the last year and half that I can go about any place and call it home almost. Of course I like to have a place that I can <u>really</u> call home, but you know me well enough to know that I am not very sentimental about such things and to tell you the truth, I don't think it is really such a good idea to be living so close to Jill and her folks. What I mean is if I should ever get married I sure would not settle down across the street fro her mother. Things like that just don't agree with me. Do you know what I mean? If you did buy that place which of course is in a very good location and all that, but it would be Bert who would live there if anybody, not me, as I said, this just wouldn't work out for me. Now look, I want you to do which you think is the best as you know more about that than I do. What ever you do decide on I will do my best to tray and help you out and I guess Bert will too.

Tomorrow I have to go on another field problem with the communications platoon and the fellows in Demolitions and Intelligence Platoons are going to jump tomorrow and as soon as it can be arranged I will be making another jump right away soon now. I hope I will be as lucky as the other times. Yes you have to be rugged. I will be spending Thanksgiving in a foxhole eating "K" rations.

It is kind of late now. I was on K.P. from 6:00 A.M. this morning until 7:00 P.M. and then from 7 to 9:00 P.M. I walk up and down the company area with full pack because I was late getting back from that 3 day pass. Oh well, a couple of nights or so and I won't have to do that anymore.

I haven't really got anything to say right now. I guess I am just sleepy. I wish I was back in New York. About the house again, I want you to get which ever one you folks want as you know which one you want and as I said, I'm not very sentimental about such things so let me know what you do.

Mom, those wings mean more to me that, than to be passing them out to relatives. I'll see if I can't get some for Grandma, but not for anybody else. I just want certain people to wear my wings and I don't think they would really mean much to anyone else. Know what I mean, okay.

I'll try to have more to say next letter, but know I had better get to bed as I have to bet up early tomorrow. Love, Bud

In the next letter dated November 28, 1943, Bud talks about receiving a "Code Machine." The code machine was made by Bud's father; Bert Sr. He took a reel to reel tape player, built a wooden box one foot by one foot, and put the reel to reel tape player in it. On the tapes were Morse code signals. Bud would listen to the Morse code tapes and practice his skills learning how to write Morse code.

Letter to Folks from Harland "Bud" Curtis

Camp Mackall, North Carolina
Letter dated November 28, 1943, Postmarked November 29, 1943

Dear Folks,

It is Sunday night about 11:00 P.M. and I just came back from the show. I stayed in Camp this weekend. I think it is the first time I have stayed in camp over a weekend since I first got a pass at Camp Toccoa, Georgia. Gosh, I slept so much today it was like I use to do when I had a morning paper route. Remember?

Thanks a lot for that code machine. It is really super and I hope to get a lot of good out of it. Boy, I sure am driving them crazy with all the stuff I got when we have an inspection. How did you finally decide about houses and all that?

Just about all last week I was out on field problems. We would get up at two or three in the morning and take off some place 10 to 15 miles out into the woods and then come in late at night and then do the same thing the next day. The only thing really hard about it is it is so darn cold and so early in the morning. My feet would get so darn cold I think I could have stuck them in a refrigerator to get them warm. These boots are all right, but when they get cold they stay cold. We were wading around through a swampy jungle one morning. I guess I'll rename this place, Quadal Mackall. It was a lot of fun but, Burrrrrrrrr, this weather and I just don't get along very good together.

How is Bert making out with the Merchant Marines? Did he quit out there at Bethlehem Ship Yards, or is he still working there?

Well I get paid Tuesday and here is Christmas coming. I still don't know what I am going to do about presents and that unless I can get out of here early next Saturday and get up to Charlotte to do a little Christmas shopping. I wish you could answer this letter before Saturday, and let me know if you think I should do the best I can or would you rather I just send the money home and you do what you want? I wonder if maybe I should just send Jill the money and tell her to get what she wants or isn't that etiquette (you know that stuff Emily Post writes). Darn her anyway. She has just about everything now.

How are those pictures coming along? Have you heard anymore from them? Let that girl across the street know if she doesn't send me a picture pretty soon now that I will get one from that brown eyed girl in New York. No, you better not tell that. I'm just kidding, but I would like to have a good picture of her and a few snapshots of Jill.

Well I guess I will be sounding off "Geronimo" again sometime this week as I take a high dive over one of these North Carolina farms. I wonder how these farmers like a bunch of guys coming down from nowhere into his freshly plowed fields. I hope there are no trees around this time. That last jump I about got hooked up on a great big one. I guess if I had of I would still be setting up there with the birds.

Well wish me luck and stuff and I'll let you know how I come out. I don't know yet what day it will be, but it ain't tomorrow as I am going to get promoted to a "Kitchen Police."

Bye for now and thanks again millions for that wonderful super code machine.

Love from a Paratrooper at Quadal Mackall, Bud

P.S. For Christmas I want a high turtle neck sweater, that's all.

When Bud had some extra money he would call home. Phone calls were about 50 cents minutes and Bud would call at nighttime when rates were cheaper and he would sometimes call collect. Bud also sent many telegrams home to his parents. The average cost was about $5.00. The telegram could not exceed more than 10 lines including the address.

Letter to Mom and Dad from Harland "Bud" Curtis

Camp Mackall, North Carolina
Post marked December 7, 1943
Sunday Night

Dear Mom and Dad,

Gosh, it was sure good to hear you again. I am feeling okay now and now I have to go out on a field problem and operate a radio for an umpire. He runs around and sees that everything goes the way it should. It is not going to be too bad, but I don't see just how I am ever going to be able to get Jill what I want if I can't get to town pretty soon. They have it figured out that we won't be back until next Sunday, so that will mess up another weekend. There is a chance that we might come Wednesday if so I will chase up to Charlotte and get this Christmas shopping done, but then there is another chance that I might not come back until next Sunday, and then get a 3 day pass and then I will go up to New York and get her something there. But then there is still another chance that we won't come in until Sunday, and then not get a 3 day pass or nothing at all until the next week and if that happens I am afraid you will have to do my Christmas shopping for me. But don't do anything until you hear from me again as anything can happen. I might not get a chance to write to you again until next week, but if I get a chance before then I will write and let you know what they have decided to do. Boy, they sure have been messing up my weekend passes and night passes with these darn field problems. I haven't been out of here since I got back from New York!

We are going up to Fort Bragg; about 40 miles away to have this problem. I will ride up there in a truck and they are going to carry our bed roles on trucks so I am really going to sleep warm this time. I have that sleeping bag Dean had while on maneuvers. Two wool blankets, one comforter (that's like a heavy quilt) and then I'll wrap all this up in a canvas shelter half (a shelter half is one half of an Army pup tent). I might even take a pillow. I'm not going to have to carry it around so I might as well take everything and be comfortable. You see this isn't our problem. It is the 3rd Battalion and we are just helping them out by operating the radios for their umpires.

Mom, I don't want you to get me anything for Christmas that will take up any room, because I will just have to send it back soon. I am going to have to send my radio and code machine home pretty soon as I am afraid we will be going on maneuvers pretty soon. I have stopped trying to figure out what we are going to do next, but I know they have something planned. Don't plan too much on me getting another furlough. Just hope for the best. Nothing this 517th does surprises me anymore so I'll just have to wait and see what happens.

Will write again soon as I can.

Love Bud

P.S. How about a high turtle neck sweater for Christmas and a wool cap. I've got everything I need now. This watch and code machine is my Christmas so put this Christmas into the house you're buying and as soon as I can get organized I'll begin sending money home to help.

P.S.S. I have seen Dean a couple of times and he is just waiting to be shipped out.

Letter to Mom from Harland "Bud" Curtis

Camp Mackall, North Carolina
Letter written December 14, 1943, the day after Bud's 19th birthday
Tuesday Night

Dear Mom,

I guess it is about time that I wrote again. I have been living out in the woods for so long I feel like a wild man.... I think I will send home a K ration and see if you like them.

We just eat them the first day and the next few days we get C rations which are a little better and if you're out a long time you get D rations. I haven't had any of those yet! I lost a little weight, but that is to be expected, but I'll get it back now. Last week was just about one of the roughest I've ever had for a long time. I was in the field 6 days last week and I must have walked close to a 100 miles or more in that time. I have really been beating myself to death, but I feel good in spite of it. You know me! I went to Charlotte again yesterday. I got a couple of things for Jill and I have been expecting to get that piano music box and pictures, but here it is the 14th (of December) *and no sign of them yet so if they do get here it will be way after Christmas before Jill will get them, but she will get these other things. There really isn't much that you can get here in the South, and it just about drove me crazy trying to find something and just about gave up in disgust. I thought you would rather I send the money than trying to find something that would probably have just been a waste of money anyway. I will send more money home next payday as I want to help out all I can on that house and I know you can use whatever I can send, so I will do my best.*

Whatever Jill sent for my birthday I haven't seen that yet either. I guess this mail service is all messed up! I will have to be sending home the radio and code machine soon now as we will be going on maneuvers soon and will it be cold. It snowed here just a little bit today and went right away.

If I can't get those things to Jill on time explain to her that if she will be patient that they will eventually get there. I don't know, but I think I am beginning to lose interest in Jill. Sometimes it goes a week or 10 days that I don't hear from her and if she doesn't think anymore of me that that, well to heck with her!

I think you know me good enough to know that I am not going to let any girl make a sucker out of me. It is hard enough to take the things I do in this outfit to have women troubles along with it. It takes a lot to keep your spirits up with things you have to do as a paratrooper, and whenever I begin to lose my morale that is going to be bad. I thought it would be a good idea to have a nice girl to look forward to coming home to, but I am beginning to have my doubts if Jill is the one. If she keeps on the way she has, I hope you will understand if I write sometimes and tell her to go to and stay there. I think more of being a good soldier and paratrooper to let a girl break my spirit. I've got this jump coming up this week and when it comes to jumping out of airplanes and knowing that you can be hurt or killed as easy as the next fellow it takes a lot of guts. When I first came into the paratroops I didn't care what happened, but when I came back from furlough I felt as if I had something worth living for, and I am darn glad that I didn't have any jumps coming up when I got back as it would have been the hardest one I would ever make. Now I don't feel that way and I don't care what happens just like when I first started. I guess that is the best way to feel because I wasn't particularly scared on the jumps I have made so far and I know this one this week won't bother me either. I'm not going to be on this side much longer and I have decided things are a lot easier to take if your nerves and feelings are hard. Then to have some girl softening them up and that furlough just about did throw me because I had such a darn good time at home. It couldn't have been better, but when I came back I was one hell of a soldier. I was thinking about home all the time and of Jill, but I don't feel that way anymore, and if I get another furlough before I go over seas I am not going to come home as I would be the same way when I came back. I guess you can't understand that or anything I have said but I will not be home again until I can come home for good. I think I will write Jill and tell her everything is all off and to forget all about me because there is nothing that makes you more home sick and softens you up like a darn girl and I am sure glad I have gotten over it now. It will take a lot to get me interested in any girl again, and if I am going to be a soldier I'm going to be a good one and the Army and girls don't MIX. Maybe I will feel completely different about everything I have said by tomorrow. I want to know what you think about what I have said and I won't say anything to Jill until I hear from you. I might even change my mind about everything by then anyway. I haven't had any sleep for 32 hours now so maybe I can blame it on that or maybe I am just trying to talk myself into something or out of something. I don't know....

I'm going to bed now and don't worry about anything I've said. Maybe this jump will settle everything anyway. You never know what is going to happen! I really don't care any way.

By now.

Love Bud

Bud and the other men were now highly qualified airborne paratroopers. The mentalities of these combat infantry men were trained to kill and to have no remorse. In speaking with Major Don Fraser, Executive Officer of the First Battalion at the 517th PRCT reunion in Portland Oregon on July 20, 2006, he said, "We were training men to be highly skilled killers to be the toughest men who had no feelings about anything." These men could not afford to be sentimental and to have feelings about family or girl friends. It would be a sign of weakness. Not characteristics of a combat paratrooper. World War II Army paratroopers were the elite of the military. The toughest our country could offer to fight a terrible war. Their training was grueling and demanding, as you have read. Bud's feelings during initial training were just that he didn't care if he made it home again. Then he described in his letter about coming home on furlough, and how great it was to be home. Bud began to think he had something to live for. He allowed his personal feelings to resurface about his love for his family and his girl friend. This was too much to take back with him to training in North Carolina. In his letter home he acknowledges coming home softened him up, and the Army and girls don't mix. Bud had to retrain his mind that he was now a combat killer. Bud could not continue to allow the feelings of love for his family and girl friend master his mind. In fact he said in his letter he would not come home again because it was too tough on him to see family and friends. These feelings had to be tucked away for when he came home from war if he could make it back. The odds were not good. If he allowed these feeling to rule him it would surely get him killed in combat. He had to be a paratrooper who took no quarter and gave none. There is an old saying in the military that goes like this: "If we wanted you to have a family or wife we would have issued you one."

In the next few letters you will read that Bud is stuck in camp pretty much by himself while the other men who live in the Southeast go home for Christmas and New Years. Training has shut down for the holidays, but Bud can't go home to California. He has just taken his furlough in October of 1943, and there would be no way the Army would allow him to go back home now. The Christmas of 1943 would be long, lonely, and bleak for Bud as you will read in his lengthy letters home during this time of year.

Letter to Mom from Harland "Bud" Curtis

Camp Mackall, North Carolina
Letter written December 24, 1943, Post marked December 26th

Dear Mom,

It was sure nice to hear your's, Dad, Bert's and Jill's voices last night even though I was half asleep from staying up to get that call through. It was almost 3:00 A.M. here when I was talking to you so I had to do some fast sleeping when I finally got back to the barracks!

As for Jill, well I'm still not sure about things. She doesn't write very much and I know darn well that she had more time to write than I do, but I always manage to find time somewhere. I haven't been out of camp on week nights since I have been back and if you have noticed the only place I have gone on weekend passes is to Charlotte and that was to do Christmas shopping trying to find something for Jill. There wasn't a darn thing there anyway that she would have liked so I got a sterling silver bracelet and had some wings put on it. I didn't get to see how it turned out as they said they would get it done and mail it for me in time so that it would get there by Christmas. I saw so much perfume that I finally got tired of looking and got some of it along with some other feminine junk. I don't know if it was any good, but it smelled pretty so I got that. I saw a music box for $15.00 that had some more feminine powder and junk and I would have gotten that too but, your letter said you were going to send me that piano box one, so I didn't get it.

Well I haven't been up there since and it is too late to get it now so she will just have to be satisfied with that bracelet and the other box of junk I got as I tried my best and maybe I can make up for it some other time. I'll send those pictures to her tomorrow as soon as I get them wrapped up. I've got all day off tomorrow! I could leave tonight (Friday) at 5:30 P.M. and I wouldn't have to be back until 6:00 A.M. Monday, but I think I will just stay here in camp. I haven't been going out much at all lately because I want to save money and send it home so you can put it on the house or maybe some day I'll have enough to get me a good car when I get out of here. I sure miss not having a car and when I get out of this Army I want to have the best one there is so I can make up for lost time then. Another reason I don't go out much anymore is because when I do I just meet some nice girl and then when I think of Jill my conscience brothers me, and I figure the money I spend now I would rather have it for something when I get out so I will send as much as I can and you do what ever you want with it and that way I know I won't spend it or get it stolen, and if I ever need any which I don't think would ever be much, unless it was a furlough I can always telegram home so when I get paid next week I'll send as much as I can. I might decide to go somewhere New Years as we are scheduled to go on maneuvers the 6th of January (1944). I am going to send my radio and code machine and suitcase home soon now. From all rumors that I hear maneuvers aren't going to be very long and then next it will be the P.O.E. (port of embarkation) and then overseas. I figure with a little stalling around and maneuvers will take up the month of January, and then some more stalling around and then getting to the P.O.E. and maybe some where in there a 15 day furlough (but don't plan on that) that will take up the month of February and so from my own deductions we should be taking that boat ride the last of February or the first of March. From the rest of my deductions by June there are going to be some of the 517th, six feet under as we are headed for combat, and you might as well know it now. You probably knew that anyway, but don't get worried about it now because it is still sometime away, and I had better take the opportunity to say what I want before censors are reading my mail!

All they are doing now is getting a few of the finishing touches put on the 517th. And getting us ready for combat. They had better ship us over soon as it is the best outfit this U.S. has ever put out. That can be proven by records of what it has done on all the physical tests, and problems, and such. When the first battalion went to jump school at Fort Benning they beat all records with 2 men refusing to jump, and then the second battalion that I was with at Fort Benning beat that record by not having a single person refuse to jump or get washed out and you can't beat that record, but now the fellows morale is pretty low (mine too) and they don't care much about anything.

There are a lot of men over the hill now and the stockade is filled up now. 8 men were AWOL (absent without leave) from this company this morning and in Headquarters Company, 2nd Battalion the whole company got together and they all went over the hill for Christmas. There are only 8 men that stayed out of about 160. They'll be back but they can't put the whole company in the stockade! Someone had better do something or they aren't going to have anyone left at all. When a man knows he is going to go into combat soon and *might not ever come back, and this might be his last Christmas, he sure doesn't care about what is going to happen if he goes AWOL, and I don't blame them. Probably if I was closer to home I would be AWOL too!*

On January 30, 2005, Bud explained that Regimental Headquarters was attached to the Second Battalion for parachute training back in 1943. He said everyone was worried about the men of the Third Battalion making all of their jumps, but everyone did okay. The 517th Regiment set a new record at Fort Benning for men completing parachute training. Bud further explained that he got to the 517th in April 1943. It was six long months for him and the men in the second battalion preparing for jump school. Men in the first battalion started their training sooner, and went to Fort Benning about one month before the second battalion. The training was rough and demanding in preparation for jump school. By Christmas 1943 everyone was a qualified paratrooper. Bud remembered most troopers were pretty well fed up with the harshness of the leaders. Everyone was now a qualified paratrooper and had nothing else to prove except going into combat. Threats of being put in the Stockade did not scare these men. They were in the roughest, toughest outfit there was. What could be done to them now?

The next paragraph Bud describes another jump he made at Camp Mackall. This jump was filmed by Hollywood, and used in the 1945 movie entitled "Operation Burma", starring Errol Flynn, the big movie star of the 1930's and 40's. Bud said it was windy, dangerous and upset some of the troopers that they were used to make this jump for Hollywood.

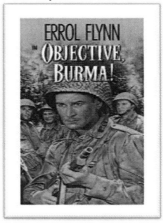

The cover picture from Operation Burma.

I have been thinking of just staying around the barracks for the next two days and catch up on some rest, but there are going to be a lot of details flying around, and there aren't going to be very many fellows around and I think maybe I had better take off out of here to be on the safe side. If I do I will send a night letter from where I'm at, but if you get this letter, and nothing else, you will know I decided not to go.

As you can see, I am still around after that jump last Tuesday. We jumped at 400 feet (the lowest yet) and it was less than 15 seconds until I was on the ground. Boy did I ever come down fast. For a while I began to wonder if I had a chute even while I was looking at it and when I looked at the ground it was about like going 20 M.P.H. in a car and looking at the ground directly below. That will give you some idea how it was coming at me. The only thing I could do is prepare to land, get ready to tumble and hope for the best. I hit hard and I tried to make a front tumble, but there was a strong wind and I was yanked out face first into the dirt, and plowed up about 50 yards of dirt before I could collapse the wind out of my parachute. The sun had been out bright and shinny all day, and the ground was soft and sandy so I wasn't as much as scratched. They asked for 16 volunteers for a night jump yesterday (Thursday) and you know me, I was the first one to step out and get my name signed up. This jumping maybe dangerous and even if it does scare me a little, I would never admit it to myself. As for jumping out of an airplane, well that doesn't even bother me. It is just thinking what could happen so if you don't think about it there is nothing to worry about. I really haven't got a good explanation for it, but it is just not in me not to volunteer for something dangerous. It might scare me, but it can't get me to be afraid, because I'm not afraid to do anything, but still be scared to death! There is a difference between being scared and afraid. You figure it out, it is too deep for me.

Gosh, but this doesn't seen like Christmas. I'm going to stop now. It is about time!

Merry Christmas and stuff.

P.S. Thanks for the swell birthday package.

P.S. Incidentally, that night jump was called off because of 50 mph wind upstairs.

Love Bud -

Bud continues on a fifth page of his letter:

I was going to stop writing because you will probably never be able to translate what I already have on the other 4 pages, but there is something else I wanted to say!

It is a few hours later now and I decided to stay in camp after all. I don't know why I am staying in. I've got a pass and I have money, but I just don't feel like going anywhere. I don't like to admit it but I think I must be homesick. I'm not very happy for Christmas eve, so that must be the reason it doesn't seen like Christmas, but when I hear all the pretty Christmas music on the radio, I'm not very happy about anything.

I hope I get a furlough before I go into combat which won't be long now. I got a V-mail letter from Garth today. He is someplace halfway around the world. I wish there had been someway we could have stayed together. He is the only guy I really like and could call a friend. I'll never forget all the good times we had in San Francisco, and Eddie too. I don't know where Ed is at now. The last time I heard from him he was in Hawaii. I guess Dean Hildreath is on his way over by now or maybe he is still in the P.O.E. (Point of Embarkation). Just a year ago tonight I was in Salt Lake City, and here I am tonight and I hate to think where I might be next Christmas Eve! (It would be Belgium and the Battle of the Bulge).

I hope Bert was home for tonight and tomorrow. What is the deal he has? Is he going to ship out or what? Well I am sorry I didn't get to send him anything, but I thought he was going to be in Washington or someplace and I thought you and Dad would rather have money. I hope everyone has a Merry Christmas, and maybe sometime I can make up for not sending Jill so very much, but I did the best I could anyway.

Good night now (again, Love Bud,

P.S. Can you send my fur lined gloves before I go on maneuvers?

This is part the 517th Christmas card Bud sent home to his Mom and Dad, December 1943

Chapter 7 - Return to Camp Mackall

Letter to Mom from Harland "Bud" Curtis

Camp Mackall, North Carolina
Letter written December 27, 1943

Dear Mom,

I am going to have to make this a short letter because it is quite late and it is time I should be in bed. I got your two letters today and after reading them I came to the conclusion I must have said quite a bit in that letter to gotten you to make telephone calls and telegrams and such. I'm sorry if I caused a commotion or something. I didn't mean to do that. There is something in the wind, but I don't know just what it is yet. Something is going to happen soon though. The only thing I know definite is that the Colonel said that we would leave on maneuvers the 6th of January 1944. How long they will be I don't know and where we will go I don't know and for all I know anything can happen so as soon as it does happen, I'll let you know.

I am still hoping for another furlough before I go to the P.O.E. and there are possibilities of it, so hope for the best. I am afraid the 517th is going to be restricted for New Years, but if not I am going someplace for the weekend because I'm afraid it will be sometime before I get another pass. I had the quietest Christmas in my life. I stayed right in camp and had a swell Christmas dinner here, but outside of that there was nothing. I am getting paid this week and with what money I have loaned out (ha, for once people owe me money) which adds up to about $45.00 and I'll get paid $100.00 minus $6.75 for insurance, $3.75 for bonds, $1.50 for laundry. I am going to send about $100.00 home and use it for what you want. Maybe I'll need it for a furlough, and if I do get home again I am either going to be engaged to Jill or there had better be a good reason why not so I will be sending all the money I can home from now on as I'm pretty sure I'll need it much more later than I do now. That is why I haven't been going out on pass much lately. Thanks again for everything. I haven't got those Christmas packages yet, but I'll probably get them tomorrow. I'll let you know when I do!

Let me know how Bert is making out. Is Lorrain going to stay with you now?
I've got to sign off now and hit the hay.

Love as always,   Bud

In Bud's letter dated December 29, 1943 he talked about the rumors of a parachute jump and maneuvers just after New Years. Archer (1985) in Paratrooper Odyssey described that, "The newly established Airborne Center at Camp Mackall was now prescribing training doctrine. One of its requirements for parachuting units was a 300 mile flight, followed by a night jump and a three-day ground exercise." On New Years Eve, December 31, 1943, the men of the 517th were loaded on 2 ½ ton trucks and drove from Camp Mackall to Fort Bragg, a short distance away. There they boarded C-47's in full paratrooper combat gear and were flown around for hours. Bud related, "We were flown down to Florida and then back to North Carolina where they made a jump." Bud feels the reason they flew around so long was to prepare for a flight into combat. As they stood up, hooked up, and began jumping out of the plane a paratrooper in front of Bud by a few individuals jumped out the door as everyone was doing. His static line was caught up underneath his mussett bag and prevented his chute from opening. He was killed on impact. This shook everyone up since it was their first death. Once on the ground field maneuvers began, and the weather turned cold with wind and rain. Bud was very cold and wondering how he was going to keep warm. The medics came into his area and told him and his friend Bill to lie down on a stretcher and to act as simulated casualties. They were taken to a nice warm tent where the doctors were acting out the part of their simulated injuries when real casualties were brought in. Bud and Bill were told to get out and go back to their companies. Once again they were back in the cold. They started walking, and unknown to them they had walked right through the opposing "Blue Army's" lines. These men were cold and miserable like everyone else, and they never stopped Bud or Bill. They kept walking down a dirt road and finally came to a small town. They saw a house that appeared to have nobody home. They checked the house and verified that nobody was there. Then they began checking to see if any windows were unlocked. Sure enough they found an unlocked window and entered the home. The house was vacant and so they stayed there the night. The next day they made their way back to their unit. No one notice they were gone. Leaders of his company thought they were in the field aide station all night.

Letter to Mom from Harland "Bud" Curtis

Camp Mackall, North Carolina
Letter written December 29, 1943
Wednesday

Dear Mom,

It is noon time here Wednesday, and I'll see if I can't get this letter started anyway before I have to fall out for the afternoon. Well I finally found out a little bit of what is going on around here. Sometime between 6:00 P.M. this coming Sunday and 6:00 A.M. Monday the whole 17th Airborne Division, that includes the 517th, and the Gliders troops will take off by truck in different directions going at least 300 miles away from Camp Mackall to different airports. Some of us to South Carolina, and some to Georgia, etc., and so as yet I don't know exactly where I am going, but it will be 300 miles from here, and then when we get there, which will be sometime Monday we will stay at a closed area to an airport. I don't know if we will get to stay in barracks there at the airport or if we will stay out in the woods eating K and C rations, but anyway when we get to the airport, the one 300 miles away, we will stay there until Thursday evening, which will be the 6th of January, and will load up our planes sometime in the evening so that we can fly the 300 miles back here to a jump field close to Camp Mackall, and we are to jump at 2200 hours (10:00 P.M.) Thursday night.

So when Thursday, the 6th of January comes around and it is 7 o'clock P.M. your time, I will be stepping out of an airplane into pitch back darkness. After we get on the ground and get organized we move to the assembly area and move up to our Camp Mackall airport, and dig in and wait until 0500 Friday morning. That will be 5:00 A.M. (2:00 A.M. your time) and then we attack this airport with blank ammunition, and by noon Friday we are to have taken the airport and glider troops come in with new supplies and then take over our position at the airport. We then move up about 30 miles to a town here by Camp Mackall and from then on we are in a defensive position and we block all enemy coming in. We stay there until the problem is over with which will be either Saturday or maybe longer. I don't know. This is going to be a mock war and will last from 6 to 8 days starting when we leave here this Sunday. I won't be able to write to you during that time so you will know where I'm at anyway.

Another thing is that February the first, the 517th will leave Camp Mackall for Tennessee maneuver area, and will not return to Camp Mackall. That was how the Captain read it to us from an order from Headquarters. So that means we will go on maneuvers in Tennessee, and will leave here February 1, 1944, and it will be the last time I ever see the place again because when we finish maneuvers we undoubtedly will go to a P.O.E. (Point of embarkation) and the P.O.E. could be anywhere. New York, or even San Francisco. We won't be told that and I doubt if any of us will know for sure until we actually get there. I'm still hoping for another furlough after maneuvers.

A few hours have passed by now and it is 4 o clock in the afternoon. I sure hope I get that package with the wool cap in it before I leave here Sunday as I am sure going to need it. Did you mean in your letter that Jill was sending me another picture? She did send me one

in black and white, but you said something about this one being tinted. I sure hope she did send me another one.

Ask Dad if he could buy me a 25 colt automatic or a 38 automatic if it is not too expensive. You see it isn't going to be long now until I am overseas, and I can have any weapon of my own that I want when I get over there. If he did send me some kind of automatic now I would have to turn it into the C.O. (Commanding Officer) and he gives me a receipt for it. The he has it shipped across with the rest of the equipment and then I could have it when I get over there and you can never tell there might be a time when a small gun that I could carry easily, and keep for special protection might come in awfully handy. Of course I'm talking about when I go into combat and I would save it for when I didn't have anything else to use. There is nothing like having another ace up your sleeve as the saying goes. So if Dad can get one and have it registered in my name and if you want you could have my name and serial number engraved on it. Ask him if he could do that will you? I was going to ask him if he could get me a good knife, but I found out that we are going to be issued a special combat knife before going across and also soon now we are going to be taught knife fighting. You can see there getting pretty serious now about our going across.

On this 8 day mock war that we are having next week we re suppose to have only one blanket, but I'm going to fox'em. I'm going to figure out someway to take my sleeping bag (I bought Dean Hildreath's the one he had on his maneuvers) with me. All I have to do is figure out a way I can fix it small enough so I can jump with it on. That jump I said I was to make Thursday, the 6th of January. I might have to tie it onto my leg, but I'm going to take it with me if I have to jump holding it over my head for my parachute (joke). It is going to be awfully cold out there for just one blanket. Heck, it would be cold if you had a dozen of them! The sleeping bag is big enough for 2 guys so maybe I'll get some one to take 2 blankets (mine and his), and I'll take the sleeping bag and then we could both get in the darn thing. With two blankets and the sleeping bag I am going to be a lot better off than anybody else out there. I think after we come back there will be a lot of fellows buying themselves a sleeping bag. I hope to have it on maneuvers too. I sure hope so because then is when I'll really need it. It will be that much more weight to carry around, but oh boy will it be worth it come night time!

I wonder if you can find those fur-lined gloves I use to have and mail them to me before February 1, 1944? Another thing I am going to have to do is send all this stuff I have back home. I am going to send it C.O.D. (collect on delivery), but I'll send home money to take care of it.

There might be a chance that I will get a 3 day pass when I come back from this 8 day maneuver. If so I am taking off for New York again. I sure like that place, and the only chance I could go is when I get a 3 day pass. So far I've only had one since I've been in the Army so I like to take advantage of them, and go someplace worthwhile. I am still going to send all the money I can spare home, and if I should get this 3 day pass I'll send you a telegram, and if you can spare it you could send me the money then. I won't be doing much New Years. Probably I'll be right here in the barrack like I was for Christmas, as we can't get a weekend pass because we will be leaving for that 300 mile trip sometime Sunday night.

Don't tell Jill this, but I sent her a K ration all wrapped up so wait until she gets it and see what she says. One of those K rations is for one meal. You have 3 for the whole day. C rations are much better, they come in cans about the size of a dog food can or middle size can of milk. There are 2 cans to a meal - 6 for the day. One can has meat and vegetables. The other can has hard crackers, hard candy, a beverage, coffee, cocoa or lemonade, etc. and sugar. That might sound like a lot, but it is not when you look at it and then try and live off the darn stuff for a week!

I mailed the pictures to you and give her the two that I signed.

I'll write again before I leave Sunday. Until then

Love as always,

Bud

Letter to Mom from Harland "Bud" Curtis

Camp Mackall, North Carolina
December 31, 1943
Letter marked on by his mother as "a very special letter"

Dear Mom,

Well today or rather tonight is New Years Eve and here I am in camp again tonight just like Christmas. I don't have the urge to travel around like I use to. I would like to get another 3 day pass before we leave here for good and I would go up to New York again. I like that place. I guess the reason I don't want to go any place is because I am trying to save money in case I get another furlough and I want to have enough to get home and maybe get a engagement ring for Jill while I am there. This is if she wanted one so it would be up to her if I got one or not. Even if I don't get home again before I go you can always use the money on the house or whatever you wanted to. Talking about money I got paid today and with what money I got and had loaned out I have the sum of $164.00 on me and a $5.00 check I'm getting cashed tonight and I still have $11.00 loaned out. Let's see that's $180.00. I'll try to send you $150.00 home tomorrow. I've got to go to the post office and get a money order made out so be looking for it soon now. That will leave me $19.00 and I might get that $11.00 before next payday and I won't be going anywhere so that will be plenty for me unless I should get a 3 day pass which I doubt anyway, but, if I do I can always send a telegram but as I said I have my doubts about getting a 3 day pass.

Hey what is this about Bert being an officer? Say that's swell. I'm glad to hear that. What rank is he? The only rating I want is "P.F.D." That's Private for the Duration. If I should ever get a chance to got to O.C.S. (officer candidate school) I would sure jump at the opportunity because I would like to be an officer, but I sure don't want any stripes. For the reason that as soon as you get to be a corporal or a sergeant you have to tell guys what to do and right away you begin to be very much disliked, but when you're an officer you don't live in the same barracks as you do when you are just a "Non Com" (noncommissioned officer or NCO) and therefore you don't associate with the officers except when you are on duty but when you are ever with the same guys like corporals etc. and you know them good and such, there is a lot of hard feelings when they begin giving you orders.

Oh, I get along with everybody fine, and so far I haven't had any trouble with any of them. The only reason I would ever want any stripes is for the extra money, but to tell you the truth I am satisfied with how things are right now. I don't like K.P. (kitchen police which is cleaning pots and pans in the dining facility) but corporals and sergeants draw other details to keep'em busy. Right now I make more (because of jump pay) than a Staff Sergeant in the regular army but I'll admit I do a lot more to get it but it makes me feel good to know that I am that much better than a regular guy in the service even though it gets pretty rough at times in the paratroopers, but there is nothing like being rugged, HuH!!! And as I said, I know that I am better than a darn high percentage of any other guy in the service; anyway that is why I got into the paratroopers, and we are suppose to be the cream of the American man and that is mentally as well as physically. Every guy in here has a mental I.Q. of a least 100 and I am in this Regimental Headquarters, and we are the brains of who is in the regiment, so I am really with the best guys you could ever find anywhere because they have to be just a little more smarter to get into Regimental Headquarters, and also the same high physical requirements. As long as I stay in the paratroopers I will never be able to go to O.C.S. though because I would have to transfer out of the paratroopers and the only way *you can get out is to refuse to jump or just get kicked out or be disqualified by some medical reason. When I get out of this army, when this war is over, I am going back to school and take everything that is required to be an officer in the Merchant Marines. I guess it would be some kind of a deal like Bob Douglas has right now. I wish I had done that and was right with him now because really I am just wasting my time here but there are a lot of things I've learned that are going to be a help later on. Well I'll tell you, the only reason I am in here now is because I was so full of adventure then and I wanted to get into something rough and dangerous and boy, I couldn't have picked a better outfit for just that thing, but now I am not quite so full of that adventure and the only reason I can figure out why I'm not is because of Jill. I feel like I have something to come home to. It gives me a different outlook on life and all that and this jumping out of airplanes and knowing I will be in combat not too long from now doesn't appeal to me like it did when I just came into the paratroopers because my chances of coming back are going to be a lot less in this outfit than it would be in some other but I am determined to stick it out now. It's just not in me to quit something when things begin to get rough and I guess if you have any faith in the religion you were taught that you will have faith enough to know that nothing is going to happen to me if it wasn't meant to be that way. If your time is up you are going to get it no matter where you are at, I guess!*

This being a paratrooper maybe rough and all that but after the way I was all my life always sick with something or another and having to go to that clinic and getting fed pills and that it makes me feel good that even in spite of all that I am now suppose to be the physically best soldier in the army and when you figure how many more men there are in the regular army than the paratroopers figures up to be a pretty high percentage.

Most of the guys in this outfit have never been sick in their life but I found out when things get going hard I am just as good if not better than they are. When we would go on some of those 10 mile runs there in Toccoa, Georgia most of the guys were hurting a lot worse than I was. There was many a run I went on where 2 or 3 guys would pass out from exhaustion, but there were darn few who just quit. They at least went on until they passed out and that takes guts. There was one run I went on where the guy in front of me dropped dead as a door nail. That was the time we were running up this 3 mile hill (Mount Curahee) about as steep as most places steeper than Shell Hill (hill in Long Beach) there on Signal Hill. It was called Mt. Curahee and it is 3 miles from the bottom to the top and it is just loose rocks all the way, and we run 5 miles before we ever got to that last 3 miles up the mountain. Well anyway there was a lieutenant that was running us, a big husky guy, and about 40 men and I was at the very tail end of all of them. Everybody went the 5 miles to the bottom of the mountain, but when we got to the top of the mountain there was only the lieutenant and me, and 2 other guys left. 10 guys passed out. One guy dropped dead and the rest just couldn't go any further. I'll admit I about passed out myself, and I was hurting so bad I actually would have felt glad if somebody had shot me and put me out of my misery, but that would have been the easy way out. Yeah things were pretty tough going those days in Toccoa when we were doing nothing but hard training and couldn't even go out on pass. Well mom as I said it got darn rough going and although we don't do any running any more there are things like these problems and this 8 day airborne maneuver I am going on this Sunday, and things will never be easy in the paratroopers but to know that I can take it makes me feel glad, even though I do my share of complaining, I would never quit. That's one thing I can say, "There has never been anything tough enough in this outfit to stop me."

I will be gone for the next 8 days starting Sunday so I won't be able to write to you so don't worry if you don't hear from me for a while. This is going to be an Airborne maneuvers just like what we will do in combat. There will be Gliders and everything. I told you that I would jump next Thursday in my last letter so you know about that. Not many of the fellows including me have made a night jump yet and we might have to make one tomorrow night so we will know a little more about it for this jump Thursday. This is quite an important affair it is only the 2nd airborne maneuver like this that has ever been had in the U.S. and it is darn important that it works just right.

I had to make a "will" out last night and I am sending it in another letter. Put them in a safe place because if anything should ever happens to me it will be a lot easier to collect that $10,000 insurance and any pay that is owed to me and all that so hang on to it. They must have something on their mind because everyone in the Regiment had to make out a "Will."

In a couple of hours it will be a New Year so HAPPY NEW YEAR and stuff!!!

Lots of love as always   Bud

Letter to Mom from Harland "Bud" Curtis

Camp Mackall, North Carolina
Monday, January 2, 1944

Dear Mom,

It is Sunday evening or rather night as it is almost 12 midnight so I won't say much. I wonder why it always is so darn late every time I write to you anyway it always seems like I start and by telling you what it time it is and that I should be in bed asleep. Well maybe I should as we are starting that 8 day maneuver, mock war, tomorrow, and I had better appreciate this warm stove and my bunk sitting right by it as I'll be sleeping on the ground for the next few days or rather nights.

Once again I was suppose to make a night jump in fact that is when I would be right now but on the way to the airport it started to rain pitch forks and stuff. It was still going strong when we got there so they called off the jump. This is the third time I have been right up to the airplane for a night jump and it was called off because of bad weather. I'll make it for sure this Thursday though because it will be a maneuver jump and maneuvers are just assimilated combat and in combat you jump no matter what kind of weather is going on out there.

You know, I still haven't got those Christmas packages you sent and I probably won't now until I get back. I'm afraid if you had anything to eat in any of them it won't be much good when I get it. Darn this mail system. It is getting a little better now though because I got a letter from Jill airmail and it only took two and half days.

I'm afraid I can't send any money home until I get back now because I haven't been any place to get a money order so don't worry now because of what I said in that other letter to be expecting it. I'll send the stuff home as soon as I get back and can get to town.

I'm going to climb into bed now so good night, and lots of love,   Bud

Letter to Mom from Harland "Bud" Curtis

Camp Mackall, North Carolina
Monday, January 11, 1944

Dear Mom,

After not being able to write to you for so long because of this maneuver I was just on, I am going to apologize for this short letter. I just wrote Jill a long one and now it is 1:00 am Tuesday morning and yes you guessed it, I have got K.P. (kitchen police) starting in about 5 hours from now but, I'll write again soon, okay!!

As for the maneuvers I had a pretty good time considering everything and as you can see I made that night jump okay.

I got your package today with the sweaters and everything and the one with the gloves and I got those films a while back with my birthday things. Thanks a lot for everything it will sure come in nice in Tennessee.

I got your letter today saying you have finally moved into the new place (3751 California Ave, Long Beach, California, the home still remains today in 2009). I sure hope I get to see it before I go across, I'm still hoping. It sounds like a really swell place.

I am sending this check made out to Dad. You see I loaned a fellow $145.00 and instead of him giving it back and then me having to get a money order I just had him write out a check. It is a good one, so be sure and cash it. I don't have very much money on hand right now, but I doubt if I will need any so just hang on to that and if it is necessary I will send home for some, but I will get along okay unless I get a 3 day pass and I would like to go to New York City, but if I don't get one (a pass) I'll just stay in camp like I have been doing.

I'm so darn sleepy I can hardly write so I am going to sign off now

Lots of Love

Bud   P.S. There is snow all over the place "brrrrrrrrrrrrr"

Letter to Mom from Harland "Bud" Curtis

Camp Mackall, North Carolina
Thursday, January 20, 1944

Dear Mom,

I got your letter, I mean telegram okay and thanks a lot because I was very much what you would say, broke and it came in very handy. I will not get a 3 day pass this weekend after all because of a big inspection I have to be here for Saturday to make sure I have all the equipment and stuff I need for this maneuver in Tennessee. There might be a chance I will get that 3 day pass the weekend after this coming one but I can't plan on as something else might come up. I would really like to go to New York again because I sure had lots of fun the last time I went there and there are so man places to see and things to do it is hard to tell you all of them. I would like to go there sometime after I get out of the Army and take my time about seeing all the places I missed but I did get to a lot of places like the Empire State Building and Rockefeller Center.

I guess I told you I jumped twice this week so far and I have 10 jumps to my credit now. More fun!!! I haven't really been doing much to speak of lately so I haven't much to say. I haven't had a letter from you all this week yet and it will be Friday in a couple of hours. I guess everything is alright though because I got the telegram. I guess that new place keeps you pretty busy doesn't it. How is Bert making out or have you hear from him lately?

Well I had better get some sleep now. I hope to hear from you tomorrow. Love Bud

Letter to Mom from Harland "Bud" Curtis

Camp Mackall, North Carolina
Monday Night, January 24, 1944

Dear Mom,

I got your two letters today they are the first ones that I got all week! I also received the 25 Colt Automatic and bullets (Pistol Bud asked his Dad to get for him so he would have personal protection while overseas). That gun is just perfect for what I would ever use it for. I can shove it right down in my boot tops along with a knife that I will be issued pretty soon. We have been having knife fighting classes lately on how to use them and sneak up on a sentry etc. I'll tell you what I would really like now if Dad could possible locate one and that is a good 45 automatic because I can get G.I. bullets for that gun and also it is a darn good weapon. Most every one in our barrack has one of those and there is going to be a time not too far off that they are going to be darn glad they have it. When I first came into the paratroops we had the 45 caliber automatic as one of our basic weapons, but the 30 caliber folding stock carbine has taken the place of that. I sent Jill some pictures where I had a carbine and I took some pictures the other day of me and one of those "Bazookas".

I fired that bazooka rocket gun the other day and it is really something. I can wear an expert medal for that now to because I got one of the highest scores on the thing. The shell that leaves the bazooka that we fired is just practice shells and they don't explode when you hit something, but they had a tank drive past at a fairly high speed and when I fired the thing I could watch metal shell go through the air and I hit e'm (tank) dead center every time. "More fun." The only explosive in the shell is too make the rocket leave the bazooka, but the real actual shell that will be used in combat have explosives in the shell that leaves the gun when it hits the tank. Well, no more tank!

To get back to that 45 – ask Dad if he knows any cops down there at the City Hall that have one he would sell as maybe he can pick one up at some second hand gun shop, or a pawn shop and get it registered in my name. The are probably a little bit expensive somewhere around $50.00, but if Dad could possibly get one for me I will sure appreciate it and if the war stays on and I stay with the 517th I can just about guarantee you that I will have plenty of need for one. I guess you are thinking, why doesn't the Army supply us with all we need? Well they really do, but as I said before there is nothing like having an extra ace to pull out of your sleeve and in other words – Don't get caught short!!! I'm sure glad I have this 25 colt, but it is hard to get bullets for it. Ask Dad if he can get some more any place. The bullets I have now will probably be all I would ever have to use for because I would only use it in case of emergency, but I don't like the idea of not being able to get anymore bullets if I ever use up those?

I like those pictures of Bert that you sent me. I hope you hear from him soon as I guess you get pretty worried about him not being able to write and all that.

I got those pictures of Jill that she had taken New Years Day, and she sure is clever. I'll have to admit that. I really like for her to send me a snap shot of her once in a while and I got a nice letter from her today. Yeah, she is okay. I think quite a bit of her and I know I couldn't hope she thinks as much of me as I do of her, and I kind of think she does. Oh well, time will tell about that… I wish there was some way to keep her from falling off garages as I would like for her to be in all one piece the next time I get home, but then I guess she just wouldn't be Jill if she wasn't doing something like that, but I'm afraid she is going to fall of one roof too many sometime and I don't know when I could find anyone else to knit me a scarf and stuff. So tell her to keep her feet on the ground for the duration, but if she still insists on climbing trees and garages, I'll see if I can't get a parachute to send to her.

Gotta go now – Good night   Lots of Love,   Bud-

Letter to Mom from Harland "Bud" Curtis

Camp Mackall, North Carolina
Friday, January 28, 1944

Dear Mom,

It is Friday afternoon and you should see this barracks. It is really messed up with clothes and equipment scattered all over. We are packing barracks bags to be shipped to Tennessee. We are packing the "B" bag with stuff we won't need in Tennessee, and that one is going to be stored someplace, and I won't see it again until after maneuvers, and the rest of the stuff goes into the "A" bag and that is what we will take to Tennessee with us. Gosh, but I have a lot of stuff to go into that "A" bag, but I'll have plenty of room in the "B" bag, I won't see it again for quite a while. Maybe I had better send this sleeping bag home. I don't know how I can get it into this "A" bag that goes to Tennessee. So if you get a package with a sleeping bag in it hang on to it, and if I can use it when I get there I'll send home for it. I don't know yet. I might find someone who has more room than I do! That other suitcase is one I bought from Dean. I bought the sleeping bag to so they are mine, but do whatever you want with them. Let me know as soon as you get the box with the code machine and the box with the radio.

  No, I didn't get the 3 day pass after all because of a big inspection we had last Saturday. This weekend we will be getting our stuff ready to move so I guess I won't see New York again for a long time unless our P.O.E. (Point of Embarkation) happens to be there and then I doubt if I could get a pass to get out of the P.O.E. I don't know nothing anymore. All I know is that we are going to Tennessee, and will be there for about 6 weeks, and that we are not coming back here to Camp Mackall. Of course I can hope for different things that might happen, but I can't plan on anything. I might get a furlough after the maneuvers is over. It is one of the things I am hoping for, and who knows, I might get it. Another thing, I might get a 3 day pass while I'm there in Tennessee, and then I can go up to Chicago. I don't know just how things are going to be there in Tennessee. If I don't get a chance to write very much it will be because I am out in the field and can't, but I'll write as often as I can.

  Gosh, but I am tired. All I have been doing for the last couple of weeks is packing and unpacking clothes from my barracks bag, and marking all my junk with my serial number. Catching up on my jumps – I have 10 now, and that is all anyone has around here. I was out on a field problem this last Wednesday and slept out. It has been surprisingly warm lately, and then sun has even been out for the last 2 days. It is suppose to be pretty cold in Tennessee. Oh well time should pass fast now with all the moving around and stuff we are doing now.

  I hope I get the box of cookies that you were talking about before we leave here. I guess I had better get back to packing again so I'll leave now and I hope I hear from you soon!

  Lots of Love, Bud

P.S. Has Dad found out anything about that 45 automatic or about if he can get bullets for the 25 automatic?

P.S. It is payday Monday and I might call up again… Do what ever you want with the money I send home.

Letter to Mom and Dad from Harland "Bud" Curtis

Camp Mackall, North Carolina
Last Day of January, Monday, 1944

Dear Mom and Dad,

  It is Monday evening and I am still here at Camp Mackall and I guess we will be leaving sometime this week or next weekend. I went out on pass this weekend and I had that urge to travel again and being as I didn't have too much money I naturally hitchhiked, and not caring particularly where I was going, I took the first car going in either direction and all together the whole weekend I went over most of the state of North Carolina and Virginia. I would just stop at one place long enough to look it over and then hit the road for the next one. It is a lot of fun just going and not caring where but the catch in it is I had to be back at reveille this morning.

  I don't think that I will jump again here in the States, the next one will be in combat unless we have a practice jump over there. I imagine they must have a regular training area or camp for paratroopers wherever it is we go to overseas and for all I know we might do a lot more training over there before we actually make the one jump that really counts over Berlin or Tokyo. I'll just have to wait and see.

We were issued combat knives today and they sure are a mean weapon.

When I was hitch hiking this weekend I got a ride in a car that was going right to San Bernardino, California and I could have gone right along with it all the way but my better judgment said no, even though I sure hated to get out of that car and see it drive away. I would have been there this coming Thursday night. Oh well maybe I will get another furlough and come home legally.

I think I will get some sleep now so write often and lots of love.

Bud

Letter to Mom from Harland "Bud" Curtis

Camp Mackall, North Carolina
Monday, Postmarked February 1, 1944

Dear Mom,

I guess you have got the night letter I sent you yesterday in Union, South Carolina. Well here I am back in Camp and it is Monday evening. I am really planning on this 3 day pass and I am quite sure I will get it and it will probably be the last chance I will get to go anywhere as we will be leaving here about the 4th or 5th of February (for Tennessee maneuvers).

I made my 9th jump today. This one was a quickie but a nice jump. A lieutenant came around and was getting fellows who had under 10 jumps and before I could turn around I was out to the Mackall airport and had a parachute on and was boarding a C-47. We jumped on a big field about 8 miles from here and at about 1,000 feet and I had a regular convention with everyone around me on the way down.

I had a rough opening shock and a couple of metal buckles hit my helmet as the parachute opened, and did my steel helmet ring. I felt like I was inside of a door bell.

I came down nice, and all in all I had a rather good time. 1700 fellows from the 517th or rather jumpers were made today. Most of them made 2 jumps today, one this morning and one when I did in the afternoon. I will make another one sometime this week and maybe even more!! – More fun…. I didn't hear of anyone really getting hurt bad today so I guess everything turned out pretty good. We have been jumping quite regular lately and I am almost beginning to feel like a paratrooper now. It's a great life if you don't weaken, but I am hurting for sleep now so I am going to call it Airborne for the day and sign off now.

Lots of Love   Bud

P.S. Can you send me some stamps?

Chapter 8
TENNESSEE MANEUVERS

In February 1944, the 517th moved to Tennessee to take part in maneuvers being conducted by Headquarters, Second Army. In the book "Paratrooper Odyssey" Archer (1985) stated, "Tennessee Maneuvers were a practice war that went on year round. Divisions and regiments came and went being assigned either to the red or blue forces. Participation in the Tennessee maneuvers was supposed to be the final test before a unit could be pronounced combat ready."

Bud describes in his next letters how Tennessee maneuvers went. He said, "No one seemed to know what was going on." At one point he is surrounded by the Blue Army, sleeps in six inches of mud in continual rain, loaded into the back of 2 ½ ton trucks at 3:00 PM and rides in the freezing cold for 14 hours. The maneuvers were disjointed and were lack ludicrous according to reports. There was no live fire and with inadequate umpire control preventing the maneuvers to be evaluated properly. Units marched day and night through rain and mud. Archer (1985) continued, "When opposing forced did meet, a sort of "Pickett's Charge" by one side or the other usually resulted." Headquarters implemented malaria discipline by having each man wear mosquito head nets. It didn't make a lot of sense because it was winter with extreme cold and rain. During maneuvers the 517th marched for 24 consecutive hours covering fifty miles. One cold day in March when all were shivering and knee deep in mud in Tennessee, it was announced that parachute elements of the 17th Airborne Division were being pulled out for overseas shipment. The 517th Parachute Infantry Regiment was told to reorganize as a Parachute Regimental Combat Team (PRCT). The news was received with great excitement and joy. They were getting out of Tennessee and going overseas!

Archer (1985) said, "Troopers of the 517th felt if there is a hell reserved for soldiers who have sinned, it must be something like the Tennessee maneuvers." Bud described this miserable experience and said, "The weather was worse than being at the Battle of the Bulge." In his letter to his Mom on her birthday, dated March 3, 1944, Bud said the following; "I will never live long enough to forget the maneuvers of 44, so now I'll have something to tell my grandchildren" (Which he has through these letters). While Bud and his sons attended the 517th PRCT reunion in Portland Oregon, July 18-21, 2006, the subject of Tennessee maneuvers once again came up in the conversation. All troopers at the reunion agreed 62 years later it was more miserable in Tennessee than at the Battle of the Bulge. The troopers stated, "You were soaking wet and cold during the Tennessee maneuvers. In Belgium you were just cold." Again at the reunion in Washington, D.C., June 29-2 July 2007, troopers stated it was miserable during Tennessee maneuvers. However, more troopers at this reunion felt the Battle of the Bulge was much colder and more miserable than Tennessee maneuvers. I think we will just have to agree that each trooper had his own opinion.

Letter to Mom from Harland "Bud" Curtis

Tennessee Maneuvers
Tuesday, Morning, February 9, 1944

Dear Mom,

Here I am sitting on a rock operating a telephone and about 50 yards down the hill is the Cumberland River. There are only about 5 of us here, the rest of the Company has moved out to a new C.P. (command post) about an hour ago. We are suppose to stay here until we get word that they got where they are going alright, but the last we have heard of them was they were having a big battle with the "Blue Army" (we are the Red Army this time) and that all of Regiment Headquarters including the Colonel has been captured so I guess we are the only ones left.

I expect any minute to have to tear down this phone and take off for the woods myself as we are surrounded by the "Blue Army."

The sun was shinning bright last Sunday and it sure felt good, but it rained all night Sunday and all day yesterday and last night. It isn't raining this morning but it has turned off pretty cold. Oh well, this is our last problem and on March 4th, we are going back to Camp Mackall. The Colonel says we won't be there long but, it had better be long enough so that we all get a furlough or every guy in the 517th will go over the hill. After all this hell I have been going through I had better get a furlough, but one way or another it won't be too much longer until I am home again.

I got Bert's letter and it sure was interesting. Tell him thanks for me. I got a card from the railway express in Nashville saying they had that gun. Now I am going to have to figure out a way to get it because I won't ever be able to go to Nashville myself. I'll write and tell them to hang on to it, and then have them ship it to me when I get to Camp Mackall, I guess.

I wish I could build a fire, but this is tactical so that's out. I'll be able to explain it much easier about what is going on when I get home than I can in a letter so I'll stop for now and write again when I get a chance to.

Love

Bud

Bud bivouacked in the woods during Tennessee maneuvers, February 1944

Letter to Mom and Dad from Harland "Bud" Curtis

Care of Postmaster, Nashville Tennessee
Wednesday Night

Dear Mom and Dad,

Well here I am in Tennessee. Boy talk about your rough nights. Last night was it. We got off the train at night into pouring rain and somewhere in Tennessee 3 miles from the border of Kentucky. I slept in about 6 inches of mud and water last night, and I was sure cold when I got up this morning. We moved to higher ground this afternoon and it is not as bad now. Right now I am lying inside my pup tent writing this by candle light.

Gosh, but it sounded good to hear you again on the telephone and thanks for having Jill there as I could talk to her. I'm sure glad to have such a swell girl and to know you folks think so much of her too. She is one girl in a million, and I'll sure be happy when this is all over and I can come home to her for good if she doesn't change her mind and still wants me. I'll just have to wait and see, I guess.

Thanks ever so much for that box of cookies, candy, and boy did that pineapple taste good and thanks for the films. I'm sure glad Dad got that 45 (pistol) I hope it didn't cost too much and when he asked me which kind of holster I wanted, I guess I wasn't thinking of the money in it as I guess it is kind of expensive.

I'm glad that you finally heard from Bert, and know he is okay. Le me know what he had to say and if he writes again. Oh well, this can't last for ever, and some day we will both be back so don't worry about me and I know Bert doesn't want you to worry about him.

Keep writing as often as you can even though my letters might be few and far between, but I will write every chance I get.

I'm going to sign off now and wrap up in a blanket and get some sleep (I hope). Tell me all about last weekend. Did you take any pictures of Jill or the house when she was there? Thanks again for everything and for taking care of my scatter brain for me (Jill).

Good night now,   Love, Bud

During Tennessee maneuvers Bud was in his pup tent with only a candle to keep warm. It was about 8:00 P.M. It was pouring rain and very miserable. The leaders of the 517th announced for everyone to break camp, and pack up their gear because they were moving out. Bud and the other men were placed in the back of 2 ½ ton trucks, and driven all over Tennessee for 2 to 3 hours, returning back to the same spot they had just left. However, now it was raining even harder than before, and the ground was a sea of mud. Bud found it was impossible to set up his pup tent again. There was nowhere to sit down. It was all mud. He had enough of these war games, and said, "I am not staying here any longer. I am going to find a place to get out of this rain. Is anyone going with me?" Two other troopers said we will go, and so they left in the middle of the night, about 10:00 P.M. These men walked about a half hour in the pouring rain when they noticed lights on at a farmhouse. Bud went to the door and knocked. The farmer and wife answered the door. Bud asked if he and the other troopers could stay in the barn. The farmer said they could not stay in the barn. He invited them into to his house, and put them up stairs in the master bedroom. The best room the farmer could offer. In the morning the farmer's wife made breakfast for all of them. The rain stopped and the sun came out. It was a large farm and after breakfast the men rode the farmer's horses. They were having such a great time they decided to stay another night. After the second day they walked back to where the 517th was bivouacked. Needless to say these troopers were in a lot of trouble. They were put on extra duty and punishment details while on maneuvers. Bud dug many holes six feet wide and 6 feet deep (6 by 6) all over Tennessee paying for his trip to the farmhouse.

Letter to Dad from Harland "Bud" Curtis

Tennessee Maneuvers
Tuesday, Postmarked February 19, 1944

Dear Dad,

I don't know how far I am going to get with this letter as we are getting ready to pull out any minute. Yesterday about 3:00 in the afternoon we boarded into trucks and we had canvas covering all the way around it so I didn't see a bit of the territory we came over, but for 14 hours we rode in that darn thing and over the roughest country there could be and was it ever cold. I don't know where I am, except they tell me we are still in Tennessee and that's about all I know. When the convoy of trucks I was with last night finally got to where we're at now, (I have that sleeping bag now) I threw that sleeping bag down on the frozen ground and crawled in, and after that truck ride it felt like a feather bed and it wasn't long until I was sound asleep. I woke up this morning and had the surprise of my life to see the sun shinning down at me so it is going to be a fairly nice day I guess, but it is still cold. Yesterday morning when I got up at the other place we were at, I found myself in about 6 inches of snow. I took a couple of pictures of it so if they turn out you can see some of the places I have been.

All around me are big hills and I guess that we are in the typical "hillbilly" country of Tennessee. I can see several old shack houses on the hills from where I am sitting now and they look just about like you see in the funny papers in that Snuffy Smith comic strip.

This is about the first chance I have gotten to answer that swell letter you wrote me and I have the general idea how the house looks from that plan you drew. It must be pretty nice.

This Tennessee is probably rather pretty in the summer time, but so far all I have seen is mud, rain and snow. I just hope this sun stays out now. A year ago if somebody had asked me if I was cold now I would have said I was pretty darn cold, but after what I have been through lately I think it is very warm right now.

I wonder what these "hillbillies" around here think about the Army running around here in their hills. Right now we are the "Blue Army" and we are chasing the "Red Army" out of somewhere around here. To tell you the truth I don't believe anybody really knows what is happening. These maneuvers here that we are on are to get us use to working with other outfits. This week we are a motorized unit and maybe next week we will be organized with a tank unit or something else.

I am pretty sure that when these maneuvers are over with that I will be getting another furlough soon. I got a V-mail letter last Sunday and I was sure surprised when I opened it and found out it was from Bert. He was on the ship somewhere around Honolulu and he seems to be doing alright. I got a letter from Dean Hildreth and he is over in North Ireland.

I am sure glad that I got to talk to you that Sunday night before I left. Thanks for getting the 45 (pistol) for me. I guess I will get it through the mail one of these days while I'm out here if you have mailed it yet. Thanks also for having Jill out there so I could talk to her to.

Well, I guess I had better stop now as I might get a chance to mail it somewhere before we pull out of here. I'll sure be glad when this is all over and can come home for good, but I'll just have to wait I guess. Got to go now.

<u>Two hours later:</u>   We have moved out of the area I was just in when I started this letter and now I am way up on top of one of these hills and I am sitting under a big tree. I can look down just like I was in an airplane into the valley where I was just at. It looks quite pretty from up here and there is a stream running through it. I think sometime before we finish these maneuvers I will slip off sometime and see what luck I have with a bent pin at catching me a fish.

This country around here is sure rocky. These hills and valley are pretty to look at from up here, but it is not so much fun walking up here loaded down with equipment and stuff.

I hear that we are supposed to walk all night. It is still early in the afternoon right now. I don't know how long we are going to stay right here. Maybe until tonight or any minute I might be moving on again.

I have got myself off away from the rest of the group so that I won't be interrupted while I am writing this. I am leaning up against a big rock and a lot of trees are all around me. I can look way over to the other side of the valley and see cows and "hillbilly" houses. They sure look small. I would take a picture of it, but I don't think it would show up and I have to ration these films.

What I wouldn't give for a big glass of milk and some of mom's cookies right now. "Oh boy." I haven't had any milk since I left Camp Mackall, over a week ago.

I can't say what or where I will be or what to expect because I don't even know from one day to the other where I will be at myself.

If things go right, the way I think they will, I should get another furlough and be home again by April anyway so hope for the best and thanks again for everything.

I'll sign off now and see if I can't find that mail orderly.

Lots of love to everybody

Bud

P.S. *This sweater and wool hood sure have come in handy out here.*

The wool hood was knitted by Bud's girl friend Jill Vignetto who made the hood with only an opening for eyes and nose, and a long tail that went down the front and back of his neck to provide added protection. Bud stated that when he was overseas, men were waiting for him to be killed so they could take that wool hooded cap for themselves especially when he was at the Battle of the Bulge. It caused some tension with others because he had it and they did not.

Bud kneeling in front with other troopers and standing in the deep woods of Tennessee during Tennessee Maneuvers

Letter to Mom from Harland "Bud" Curtis

Somewhere in Tennessee on Maneuvers
Tuesday, February 22, 1944

Dear Mom,

Well I am still in Tennessee and that's about as much as I know about where I am at. A couple of days ago I was near Murfreesboro, that's about in the middle of Tennessee. I think I am somewhere off 100 miles from there now, so you can look it up on the map.

It has been raining continuously just about all the time I have been here, and it is raining right now. I am under a raincoat operating a switch board right now and it is fairly *dry under here. I am sure getting tired of wading through this mud and slippery rocks. We are now on another problem (field training problem) and we camped right on the side of one of these hills with a lot of trees around for cover from the air. I crawled into my sleeping bag last night at the top of a hill and when I woke up this morning I was about 50 yards away at the bottom of it (more fun). Quite a few fellows have got sick from all of this bad weather and have had to go to the hospital. So far I haven't had as much as a sniffle or a sneeze. I wouldn't mind one, but if I got double pneumonia and got put in a nice warm hospital, but I'm not that lucky and besides that is the easy way out so I guess I will just have to stick it out for a while longer. I just got a letter from you so I am going to read it now and see what is going on in the land of sunshine.*

I just finished reading the letter and I am sure glad to hear that Bert is home. It is getting dark fast and I had better go get some chow before it is too late. If you want to send something, make it something to eat. I haven't got the gum or Jill's cookies yet, but I guess I will soon, as this problem is over, and we get another rest period.

Thanks for your swell letter.

P.S. *A guy just brought up another one from you but it will be dark before I can read it and we can't have lights, so I'll have to wait till tomorrow. Lots of love always.*

Bud

Bud at the bivouac site during Tennessee Maneuvers

Letter to Mom from Harland "Bud" Curtis

Tennessee Maneuvers
March 3, 1944, Mom's Birthday

Dear Mom,

It is just past 12 midnight and today is your birthday. I feel awful bad because I can't send you anything or even get to a telephone and call you up. We are leaving here Saturday the 4th, and are going back to Camp Mackall. I will call you up as soon as I can when we get there so I'll just have to wait until then.

I still have a letter I wrote to you a couple of days ago that I haven't had any place to mail as yet, so I guess you will get both of these together. That field problem is over and right now I am sitting by a raving fire (1 candle); well it's a fire anyway and it is putting out enough light so I can write this. I got your letter a little while ago with the pictures in it. Bert looks pretty darn good in his uniform! Reid Johnson looks okay too (Boyd Johnson's son). I guess the neighbors think you have a U.S.O. there or something with the two of them running in and out.

Honest mom, a couple of days ago when we were on that field problem, and I was sitting, and leaning against a rock in <u>the rain</u> shivering all the way through with no place to get out of the rain. I couldn't build a fire because of the danger of the other army seeing us and hadn't had anything to eat for 24 hours, and nothing to look forward to but the night time when it would get more colder. All I could do was sit there with my head bent between my knees and clench my hands together and grit my teeth and dig deep for a little more strength and determination to hold on. Then I would think of all the other things I wondered if I would live through. Sure "I'm a paratrooper" the most rugged soldier in the world, I can take anything (anyway that's what they tell me). I've jumped out of an airplane in the pitch dark night then walked all night and still going through the next day. I've ran until I thought if I took another step I would drop dead yet I kept on going for miles farther. I've done pushups till I fell on my face, yes, I would think of all those things, but then I would remember that there was only one reason why I have done all these things and that was pride, but the only reason I had that pride was because I was always thinking of home and you and Dad and Bert (Bud's brother). It's kind of hard to explain what I am trying to say as I just can't express myself the way I want to, but it all adds up to I came from the best family and have the best most wonderful mother in all the world and without that I couldn't have made it through anything not even the first day that I got to Toccoa, GA., let alone all the other things so far. Well that other night when I was so cold and wet is just another memory now and I am still going strong and still no sniffles or sneezing. Quite a few fellows are in the hospital, but I am still here, and I will never live long enough to forget the maneuvers of 44, so now I'll have something to tell my grandchildren.

Thanks again for your letter and I am very sorry I can't send you anything, but all my love to the best mother there is.

Lots of love and kisses, your paratrooper

Bud

XXXOOOXXX

CHAPTER 9

ONE FINAL TRIP BACK TO CAMP MACKALL

From the mud of Tennessee, the 517th had proven to the Army that it could function as its own separate Regiment, a Combat Team. The unit was quickly sent back to Camp Mackall to prepare for overseas movement. The 517th received no special augmentation except it did have artillery and engineers. The 460th Parachute Artillery Battalion and the 596th Parachute Engineer Company. The 517th was expected to be independent and operate as a small Division, although it lacked many of the specialized services that were assigned to a regular Division. The 517th was now standing alone, and would have to make do with what they had. The regiment quickly returned to Camp Mackall from Tennessee, and all efforts were concentrated in preparation for movement to the European combat theater. Wills, powers of attorney were made out, and shots were given to every trooper to prevent diseases. Crew served weapons, artillery, and vehicles were given a coat of cosmoline, crated, and readied for shipment. When all was done companies and batteries then prepared their individual weapons for combat. And yes furloughs were granted allowing the men one last trip home. Bud quickly went to Marietta, Georgia, and caught a military hop home to California.

Bud Curtis in March 1944, home from Camp Mackall, North Carolina, on his last furlough

It was unusual for a master sergeant to associate with a private however one day Master Sergeant (MSG) Dubois was asking his men for money so he could go to town on a weekend pass. No one volunteered to give MSG Dubois any money. When Bud left camp for his weekend pass he began to feel sorry for MSG Dubois. Bud described, "I went to town and found a Pawn Shop. There I got $10.00 for pawning my wristwatch. I returned to camp and found MSG Dubois. I told him here Sarge now you have $10.00 to go to town. MSG Dubois said wait a minute, no one had any money let me see your wrist. He saw my watch was gone. You pawned your watch didn't you? I said what if I did. He was overwhelmed. He could not imagine anyone doing this for him." Well, Bud said I am leaving and going back to town. MSG Dubois said could I go with you? Bud said sure. Both men went on pass to town together that weekend. Over time Bud and the MSG Dubois became close friends. MSG Dubois and Bud went on pass everywhere together. One day MSG Dubois formed a work detail. Even though Bud was his friend MSG Dubois had him on a work detail with some other troopers. Their assignment was to pickup trash from the ground, however to be just a little bit crueler MSG Dubois made the men squat down, and walk like a duck, and squawk like a duck. Every time a man picked up a piece of trash he had to holler out "QUACK!" Bud was okay with this treatment because it was given to every man on the detail, but when Bud asked MSG Dubois for permission to go to the barracks and get his gloves because it freezing cold, MSG Dubois refused to let him go. It was a simple request and Bud's hands were freezing. Bud thought this guy is supposed to be my friend. Well Bud never went on pass with him again, and would not associate with MSG Dubois thereafter.

Bud with Master Sergeant Dubois at Camp MacKall, N.C. 1944.

Letter to Mom from Harland "Bud" Curtis

Regimental Headquarters
Camp Mackall, North Carolina
Wednesday, April 5, 1944

Dear Mom,

 I finally got back here to the Army routine again and sure wish that I was back in Long Beach. I still could help on that darn motorcycle (brother Bert's motorcycle) or go lie on a warm beach. Yeah, it was really wonderful while it lasted and I am afraid it will be a long time until I will be doing that again.

 I got in here last night five days late. I was rather lucky though because this morning the Company Commander took off on his furlough and a lieutenant is taking his place while he is gone and he is a pretty good egg. He only counted me as being 4 days late which is 96 hours and all I have to do is make up that 96 hours I was late by working K.P. (kitchen police) for four hours each night and all day each Sunday until all of the time is made up. If the regular C.O. (commanding officer) was here I would probably have gotten a court martial and a heavy fine, so I really am getting off easy (I guess). That Jack Dunaway (a friend from Hollywood who Bud met in the 517th) that was out to the house last Wednesday, a week ago today, remember? Well he isn't here yet!

 If I had known it was going to take me so long to get back I would have stayed at home a couple or more days and could have made just as good of time by waiting until a plane did go from Santa Ana. I don't know though. If I had done what I intended on doing, I would still be home but everybody seemed to be worried about me not going back on time; especially Jill. I decided I wouldn't be very happy staying there anyway with everyone feeling that way.

 I took my time coming back because I was going to be late anyway. I got off the train in El Paso and stayed at an (Army) airport Thursday night (perhaps March 27, 1944?) and got a good night sleep (in a barracks). Then I caught a plane Friday morning to Dallas, Texas and the pilot was going to stay there for the night and leave at 8:00 am, Saturday morning and he told me to go on into town and be out at the airport when he left and I could go onto Jacksonville, Florida with him.

 I had a good time adventuring around Dallas until I ran into an M.P. (military police officer) and naturally my time on furlough had run out by that time and I couldn't see it his *way when he said he was going to take me in and while I was in the middle of kicking his teeth out a couple of more MP's came up from some place and I was kind of out numbered and one of the guys caught me off guard and got me from behind with a billy club or something and when I woke up I was in a guard house with an awful head ache. I had my train ticket all the way back to camp so they let me go the next morning and gave me some other papers so that I wouldn't be bothered any more, but I was too late to get on that plane to Florida. So then I decided to get back to the train route and I started hitch hiking on to Houston, Texas. I got food poisoning somewhere along the line and boy was I ever sick. I passed out along the lonely Texas highway sometime Saturday night and woke up a few hours later and felt much better and then I caught a ride on into Houston. There wasn't any train leaving until morning so I went out to an airport and they told me I could catch a plane in the morning and they gave me some bedding and the top floor of a big barracks to sleep in for the night (that was right after I called you up on the phone). I really slept good and didn't wake up until late Sunday afternoon and it was raining cats and dogs. All planes were grounded so I got some good chow (food) at the mess hall and went to a camp show and back to another good night sleep Sunday night and got up bright and early Monday morning and caught a plane to Albany, Georgia. I took a bus from there up to Atlanta Monday night and got there early Tuesday morning and caught a plane to Charleston, South Carolina and from there I hitchhiked on into camp and got here about 9:00 pm last night Tuesday (April 4, 1994). I got up this morning and had K.P. all day. Went to a show this evening and here I am sitting on my bunk writing this letter. Well, now you know where I have been and what I have done since a week ago tonight when I got on the train in L.A. (Los Angeles).*

Well you don't have to worry anymore because everything is running smoothly, and I am back to being a soldier again and I don't think I will be going anywhere for awhile so keep on writing and let me know what every one is doing. I still wish that I could have seen Jill again before I left because it will probably be a long time until I am there again but then I myself can't predict where I will be from one day until the next. Tell Jill hello and love and stuff to everyone

Love

Bud

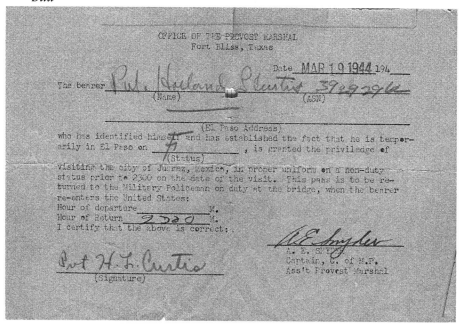

The pass Bud was given in El Paso, Texas

Navy PBY Catalina type aircraft Bud flew on from El Paso to Dallas, Texas.
PB stands for Patrol Bomber and Y was the designation of Consolidated Aircraft
Picture taken at the Air Force Museum at Wright Patterson Air Force Base, Dayton, Ohio

The following is the story Bud recalled about his trip back to Camp Mackall from his last furlough as told to his son L. Vaughn on May 9, 2004.

Bud was given furlough from Camp Mackall, North Carolina during the 3rd week of March 1944. This was his last furlough to go home before deploying overseas. The 517th PRCT gave each man only five days to see his family. How could Bud get home to California with only five days? The train took five days to get to California. The train was out of the question. Bud knew how to catch military flights from his old buddy Smitty, and so he left North Carolina for Marietta, Georgia, where there was an Army Air Corps Base. Bud was very lucky. He arrived there, and found a flight going directly to the Army Air Corps Base in Santa Ana, California. This was only 20 or so miles from his home in Long Beach. He was so proud of himself for catching this flight while many of his other friends were still traveling home on a train. Bud enjoyed every minute of his furlough.

The long awaited 45-caliber pistol he had been asking his Dad for was finally given to him. He would keep it as his secret weapon. Although he was not authorized by the Army to have this pistol, Bud felt he should have a backup gun just in case. Bud decided it would be easy to catch a flight from the Air Corps Base in Santa Ana, California back to Marietta, Georgia. On his fourth and second to last day of furlough he went to Santa Ana to catch a flight back east. There were none. Bud was shocked. I have got to get back to Camp Mackall. I have to be at reveille at 0600 hours tomorrow. What am I going to do he thought? Bud's father said to him the only thing you can do is go back by train. That would make Bud 5 days Absent without Leave (AWOL). Bud could face a court martial for being AWOL. He had no choice. His father purchased the train ticket back to North Carolina. Good byes were said, and Bud got on the train in Los Angeles, California. It seemed to take so long. The train traveled so slowly and Bud had to sit in a seat the entire way. The train headed for El Paso, Texas and arrived there early that evening. Bud decided he had enough of riding on the train and got off. He knew his furlough expired the next night at midnight. He went to the Army airfield in El Paso, and found a Navy PBY aircraft leaving for South Carolina the next morning. He made arrangements to get on this flight. That evening he wanted to go to Juarez, Mexico and visit. So he secured a pass and visited Mexico. The next morning he reported to flight operations where they gave all passengers a parachute. He got on the PBY that is a high wing, pontoon shape fuselage, twin-engine aircraft that lands in water. They took off for South Carolina, but the plane developed engine trouble over Dallas, Texas. Bud said to himself he wasn't staying on that plane if there was a chance of it crashing. He quickly moved toward the door to jump out. After all he did have a parachute on and had already made 10 parachute jumps. Jumping out of a plane did not scare him at all. He started to jump when the pilot, Lieutenant Roberts shouted back, "Don't jump, I can land this thing." Bud returned to his seat and sure enough the pilot landed the aircraft. Repairs were made, and the flight was scheduled to leave at 7:00 am the next morning.

While waiting in Dallas, Bud was allowed to stay in the barracks. He had the 45 caliber, colt automatic pistol his father had given him. He was concerned about being caught with it so he left the base, and went downtown to rent a room at the local hotel. That evening Bud left the gun in his hotel room, and went out on the town to a roller skating rink. There he met a girl one half hour before the rink closed at 11:30 pm. He convinced this girl she did not want to walk home by herself so Bud volunteered to escort her home. They hadn't gone but a few blocks when military police (MP) officers in a jeep pulled up, and demanded to see Bud's pass or furlough papers. It was 2 minutes after midnight. The MP said, "Your furlough expired 2 minutes ago, you are coming with us." Bud was put in the jeep and waved goodbye to the young lady, whom he never saw again. The MP's put Bud in the stockade (jail) cell. They told him he would have to stand before the magistrate at 8:00 A.M. the next morning. This would not work. Bud's flight back to South Carolina was leaving at 7:00 A.M. He demanded to be let out so he could make his flight time. They refused. Bud was awfully hungry and asked for food, which they gave him. Shortly after eating the food he became violently ill and very sick to his stomach. He vomited all night and did not sleep a bit. 7:00 A.M. came and went, and at 8:00 A.M, still being sick he stood before the magistrate (judge). He explained his story showing the magistrate his train ticket, then told him about his PBY flight with Lieutenant Roberts. The magistrate checked all of this information and found it to be correct. Finally, someone believed him. The magistrate ordered the MP's to take Bud back to the train station, sit with him, and ensure he got on the train for Camp Mackall. There Bud sat with two MP's sitting on either side of him. Bud was very worried about his personal belonging that he had left in the room in the hotel downtown the night before. His bag had that 45 Caliber pistol in it. He knew if he did not get to the room by check out time at 1:00 P.M. his stuff would be taken, and he would loose his pistol his father had just bought for him. Bud still had his room key in his pocket. There he sat at the train station with the two MP's. What would he do? The train was scheduled to leave at 9:00 am. The time came and no train. The MP's checked on the reason why it wasn't there. Delayed. It would be hours before it got into Dallas. As the 3 sat there, Bud told the MP's that there was no reason for them to hang around. He had nowhere to go. He had his train ticket, and assured them he would get on the train when it arrived. There they sat. Two MP's staring at Bud and waiting for the train to arrive. Bud told them they could go. No reason to hang around. They had done their duty, and he assured them he would remain at the station until it arrived. The MP's bought the story and left. Bud jumped up, and ran out of the train station for the hotel downtown. He arrived there just before check out time and got his stuff. His pistol was still there. As he left the hotel he knew he could not go back to the train station. There was no way he would arrive back to Camp Mackall in time. He knew there was an Army Airfield in Houston, Texas. He decided to hitchhike there. He got on a local Dallas city bus, and took it as far as it would go to the out skirts of town. There he got off and began hitchhiking. Gasoline was rationed then, and there were very few people traveling by car. Only trucks that had reasons to use gasoline were traveling. So a truck driver picked him up, and away they went as fast as the national speed limit of 35 miles per hour (MPH) would allow. It took Bud 24 hours to travel 300 miles to Houston. When he arrived he immediately went to the flight operations desk. He found out there was a special plane traveling the next day to pick up a general officer in South Carolina. The plane was to leave at 9:00 am sharp. Bud was taken to the transient barracks where he was given a room to stay in. He was still sick and went right to sleep. He slept until 10:00 am the next morning when he awoke to a downpour of rain. He realized he missed his flight once again. He ran as quickly as he could back to the flight operations desk to find out what he could do. They told him to relax. The flight was cancelled because of weather and to come back tomorrow. He went back to his room, and ate food for the first time in 48 hours. The next day he was over being sick and back to his normal self. At 9:00 AM he got on this plush aircraft setup for a general. He was the only passenger and away he went. They landed in South Carolina where he caught a train to North Carolina, and Camp Mackall. At that time Bud was assigned to the regimental headquarters. When he reported back in his company commander was gone on furlough. The lieutenant left in charge told him, "Curtis, you are very lucky the Captain is not here. Because you are 5 days AWOL I am not going to court martial you. I am going to put you on K.P. (kitchen police) for 5 days." Bud was very relieved and accepted his punishment. He reported for K.P. He washed pots and pans, scrubbed and peeled potatoes. He hated it. After one day of all of this K.P. he had enough! The next day after reveille he decided he wasn't going back for more K.P. He decided to hide under the barracks all day. At the evening formation he would appear. No one was wise to what he was doing. The sergeant in charge of K.P. did not know how many days Bud was suppose to be there so he did not report any thing unusual. Bud hid under the barracks for the next 4 days. He then reported for regular duty.

Letter to Mom from Harland "Bud" Curtis

Regimental Headquarters
Camp Mackall, North Carolina
Saturday, April 8, 1944

Dear Mom,

Tomorrow is Easter and I can't even get out of camp to send anything so I guess you will just have to forgive me and also I am awfully sorry I couldn't send Dad anything. I wish I could have stayed just a little bit longer.

Gosh I am the only one left here in this barracks; everyone else had taken off on a weekend pass and some are still on furlough, and then there are the ones who live pretty close to here. They got a 3 day pass to go home. Oh well, I haven't got any place in particular I could go anyway so I might as well be here. That extra K.P. (kitchen police) I have to do isn't hard, but I have to do it after duty hours from 6 pm until 10 pm and that cuts right into the middle of the whole evening.

Just 80 more hours to go now and I will be finished but the way things are going I might be finishing up on a boat as I don't think we will be here very much longer. The latest rumor seems to be that by the 15th of the month our training will be completed and by the 23rd we are to be out of here for good because a new Regiment is suppose to be moving in here and take over our area. I'll just have to wait and see how that one turns out. Its just a rumor and I don't believe anything until it actually happens.

I got some 45 ammunition for my gun and tomorrow I am going out to the range and do some practicing.

As Jill would say, "Hasta Banana" and a very happy Easter to all.   Lots of love
Bud

P.S. If you do see Jill tell here I am very sorry I couldn't send anything to her either.

Letter to Mom from Harland "Bud" Curtis

Regimental Headquarters
Camp Mackall, North Carolina
Monday, April 10, 1944

Dear Mom,

I am just going to write a short line to say I am okay and everything is running smoothly. I am going to put in the ticket I bought in Los Angeles (meaning he sent back the unused portion of his train ticket his parents bought him on his last furlough). I only used it up till El Paso, so see what you can do about getting some money back on it.

I haven't heard from you for a few days so I don't know much what is going on there. I hope Bert made out okay in San Francisco? I got a letter from Garth (Boyce) and Frank Hill and they both seem to be making out okay except they will be glad to get back here to the States. I guess things aren't so good over there. Who knows, I might be seeing them soon or be in that vicinity if our P.O.E. (Port of Embarkation) is San Francisco. I sure hope that it is Frisco, as I might get a chance to come home, but if we left from there we would go to some island and I think I would rather be in Europe and that is where we would go if our P.O.E. is New York, so I don't know which I want and anyway I can't do anything about it except hope everything turns out for the best!

Write soon and before I forget, I want to thank you for the swell box of candy you sent me. It sure tastes good.

Bye Now
Love   Bud

Last portrait taken of Bud while he was home on furlough March 1944

Chapter 10
TRANSFERRED OUT OF REGIMENTAL HEADQUARTERS

When Bud returned from his final and last furlough from California, Master Sergeant (MSG) Dubois his Section Sergeant in Regimental Headquarters was there to meet him. Bud had been snubbing MSG Dubois after the work detail incident. Bud would go on pass to the local town without his old friend. This angered MSG Dubois, and he put Bud on weekend K.P. details to get revenge for being snubbed. Now that Bud was back from California MSG Dubois had Bud declared excess in Regimental Headquarters. Bud was no longer in the 517th. PRCT. What would he do? He started asking every company and battalion if they had a spot for him. None did. It looked like he was out of his beloved 517th. For five long days Bud tried to find another position with no luck. He finally talked with Staff Sergeant Jolly from Shreveport Louisiana, the Communications Sergeant of the first battalion. SSG Jolly told Bud he only had a "runners" position in the battalion. Bud was a communications operator, and now to be demoted to runner was a bitter pill to take, but he did it. Luckily, Bud was still in the 517th. He now didn't have to leave the unit. In later years Bud remarked, "Shortly after I found a home in the first battalion MSG Dubois got into a card game, and called the Regimental Lightheavy Weight Boxing Champ, Sergeant Lemanski a cheat. Sergeant Lemanski took exception to the accusation of cheating, and beat Dubois to a bloody pulp. From then on just about every man in the outfit despised MSG Dubois for calling the Champ a cheat. Soon thereafter MSG Dubois was reduced in rank to Private, and shipped out to another company." Bud never saw Dubois again, but he figured Dubois got what he deserved.

Letter to Mom from Harland "Bud" Curtis

Camp Mackall, North Carolina
Monday, April 20, 1944
Headquarters, First Battalion

Dear Mom,

I am sitting out in the woods under a big shade tree right now. All we have been doing all week is packing equipment to be shipped. Everything is covered with cosmoline first and then packed into water proof boxes, so I guess you can get a good idea what is going on. We had to turn in all our cloths except one "Class A" uniform (This would be his dress uniform with gold buttons). When we get to the P.O.E. (Point of embarkation) we will be issued everything new. I don't know just when we will leave here, but it should be very soon. And of course I don't know where I am going, but I am afraid it will be for a long time though. I am sure glad I got to come home this last time, and as for being those few days late. Well that is all forgotten now, and as for me, I just wish that I had stayed longer.

I had a wonderful time at home being able to sleep in my own bed and as late as I wanted to. There is nothing like digging into the refrigerator and having so many nice things to eat either. All in all, even in spite of that fuss with Jill, I had a perfect time. I still don't know exactly what it was all about and I can't help wondering what the deal about that was. I was just gone a few months this last time and as soon as I came back all that happened. How will she be when I go farther away and maybe for a couple of years (who knows? I don't). The last few days I was home everything went pretty smoothly, but it seemed to me that she was just being that way because she knew I wouldn't be there long and it would be easier that way. I really had planned on getting engaged to her when I was there, but she said no, and she never mentions anything about it in her letters. I am beginning to think that I have just been kidding myself, all along.

Now that I have been back for a few weeks and have had a chance to think it all over now I have decided it must have been a mistake all along and before my mail begins to be censored I had better tell her it is all off and just to consider me as a friend only, or something. I personally think she would rather have it that way too. I might be wrong about all this, but if I am it isn't because I haven't tried to have things different so my conscience won't bother me any (and I am sure that I can find someone else who is just as nice and at least let me know what is going on and not keep me guessing all the time). Before I do say anything to her maybe I should wait and hear what you think about it. Maybe if you talked to her yourself you might be able to find out if I am wrong or not. All I know is I don't ever want to put so much trust and faith in any girl again the way I did her and come home and have her act the way Jill did the first couple of days. I never have gone with just one girl as long as I have with her, and it isn't easy to give her up just to say to heck with her, and forget all about her, but I will have plenty of time to forget everything about anything from now on. I want to get everything settled before I do go overseas, and know for sure one way or the other.

I got the watch yesterday, but the hands are real loose and slip all around. As soon as I get out of quarantine I will get it fixed in town. I mailed that train ticket to San Francisco, so that is taken care of. It is about time for Retreat (Retreat is a formal ceremony where all the army units fall out and line up in formation for the lowering of the flag). So I had better finish up now. I guess I really shouldn't be bothering you with my troubles with Jill, but you told me to, so there they are.

Lots of Love Bud

P.S. I am glad to hear Bert is getting along okay, and tell Dad thanks for writing and finding out about the train ticket

Letter to Mom from Harland "Bud" Curtis

Camp Mackall, North Carolina
Monday, April 24, 1944
Headquarters, First Battalion

Dear Mom,

The first thing I had better say before I go any further is from now on my address is Headquarters Company, First Battalion. They were very much over strength in Regimental Headquarters (meaning there were more soldiers assigned to Regimental Headquarters than there were positions to fill. Too many people in one company), and they needed communications men here in Headquarters, First Battalion, so several of us were transferred over here and other fellows to the second and third battalions. The main thing I don't like about the who deal is now I will have to make a lot of new friends, and find out who I can trust and so forth. It has its good and bad points. As for the barrack I live in now it is just behind he one I was at in Regimental Headquarters, so I am just a stone's throw to go over and visit the boys. I am only a couple of barracks up from Charles Diegh, but I haven't seen him as yet.

It has been raining off and on all day. I didn't get back from weekend pass until about ah hour before reveille this morning, so I didn't get much sleep, so I guess tonight I will hit the hay early and catch up on the shut eye. What does Bert do now that he is 1A? (1A is the draft classification for men fit to be drafted into the military) can he still get a boat or does he have to wait till the draft board gets him? If that is what does happen tell him if he has any sense at all to do everything possible to get into the Navy. I'm sure he would like it much better unless he could get into the Army Air Corps. I think I will ask for a 3 day pass tomorrow and if I get it I will go up to N.Y. again. Well, that is about everything for now. Write often. Love Bud

P.S. You hadn't better send me anything more because I don't know just how long I will be here (thanks for thinking of it though)

Letter to Mom from Harland "Bud" Curtis

Camp Mackall, North Carolina
Monday, May 1, 1944
Headquarters, First Battalion

Dear Mom,

I am sitting down at the telephone building right now and I am trying to get a call through to you. It is 12 now, and there is a 4 hour delay unless something happens so if I can't get the call through by the time I finish this letter, I am going to call it a nite and go to bed, and call you up sometime this week and then I might get a chance to send you a nite letter or something and have you drag Jill out there and talk to her too. I would sweat this call out tonight except I am so darn sleepy I can hardly write. I haven't had any sleep now for 2 nights because of traveling and stuff.

I just made it into camp this morning from New York just in time for reveille. I took in all the things that I missed the other times. Brown eyes drug me up some spiral steps inside of the Statue of Liberty and when we stopped I found myself looking out of a little window right in the very crown on top of the ole gal's head. Then she leads me through some underground tunnels to where I find myself on a underground street car heading for Coney Island. I went down on the beach and walked around the water's edge playing tag with the waves. I made a confession in a letter I wrote Jill today and told here the truth about Brown eyes. She is 23 years old and whenever I am in N.Y. she takes care of me like she was my big sister, and makes sure that I don't get lost, strayed or stolen. Then when the time comes for me to leave; she see's that I get on the right train at the right time. I would probably be still wandering around in one of those underground tunnels if it weren't for her. She is very nice and from the pictures I showed her of Jill, she thinks Jill is very cute and is sure that she would like her.

I got paid today and I had this Sergeant (Staff Sergeant James Pease) *make out another check for $120. I hope that it hasn't been too hard on you to send me that money so that I could take that 3 day pass. It was rather expensive, but it won't be long now until I will be a long way from all of this, and I am sure glad I got to go up to New York as from all I can gather, we will be in a P.O.E. (port of embarkation) by next week, unless something happens to hold us back a little longer. I will call you before I leave though. I didn't get paid too much this month because they deducted for the day I was late. I couldn't resist the temptation of selling the 45 caliber pistol and holster for the sum of $96.00* (Harland's father Bert Sr. purchased a 45 caliber pistol so he would have a personal weapon when he deployed overseas to the war)*. How much profit is that?*

I really didn't want to sell it, but I hated to pass up such a bargain, so I sold it. There is a chance I might not have gotten it through P.O.E. anyway. I am keeping the small 25 automatic because it is so easy to hide. I can take it apart and put it in my watch pocket. I have a box of bullets for it that a fellow got me from Chicago.

Anyway I am sending the money home, and I still have plenty left to see me through this month, and maybe I can send some of it home if I see I won't need it.

I made out a new deductment for bonds and next month they are suppose to start taking out $25.00 a month for bonds and every three months they will mail you a $100 bond. I had them make it out in my name and Dad's name, so anytime you want to you can always cash them in. After I get overseas and see how the situation is and where I am going to be at I might have them take out money and send it direct to you.

This headquarters, first battalion, that I am in now seems to be O.K. The fellows are pretty good eggs and you can't beat the first sergeant that I have now. He was the one who got me that 3 day pass.

I got that big box today and I will be the next 3 months eating all of the things you sent. You must have bought a whole store out and put it all in that box. Thanks for all of the nice things; they sure taste good.

After reading your letter, I am sure that Jill does love me very much and I won't worry or have any doubt about it again.

I hope that I get to see those colored films before we pull out of here. How did they turn out?

I don't like to say for sure just when we are leaving because I don't know for sure. We are suppose to be out of here by this week. All of our equipment is packed away in water proof boxes and we have been issued a completely new everything from shoe laces to tooth brush. Now we have been ordered to remove all patches or anything to identify us as paratroopers. When we move out of here, we will be wearing G.I. shoes and leggings and all the trimmings of a brand new rookie just out of a reception center. I have two brand new pairs of boots that I probably won't be allowed to wear until we get to our destination. It is 2 am now and I am going to call it a day and go back to the barracks. I will call you up sometime this week though. Maybe even before you get this letter.   Thanks again for everything. Love Bud

P.S. *Have the press* (meaning Press Telegram Newspaper from Long Beach, California) *mailed to me at Hq, 1st Battalion.*

Letter to Mom from Harland "Bud" Curtis

Camp Mackall, North Carolina
Friday, May 5, 1944, Headquarters, First Battalion

Dear Mom,

Here is some more stuff I have to get rid of. Hang on to it and maybe I will get some place where I can have it again and then you can mail it.

I hate to send Jill's picture, but I have so very little room in that duffle bag that even if I could squeeze it in, it would get all smashed up. Maybe by the time I send for it, she will have a new one for me.

Say, that gives me an idea. With some of the money I sent home and if you want to get me something I really want. Take Jill down to a good photographer and hold her still long enough to get one of those beautiful colored pictures taken, and then send it to me. I sure like the one that she has in that Bible she sent me, but I want a real big one also, so if you ever have time please do that for me will you?

If you can get a hold of some more colored film, take some more pictures of Jill, you, dad, Bert, and Lorrain. Well just take some more because I sure like the way these here turned out.

There is nothing I like better than to see pictures of all of you, even if I do have to send them home, it is a lot of fun to look at them.

If you look around enough in this box you will find a little gold box. I hope you like what is inside of it.

Well, I have a very busy day tomorrow so I will stop now.

Always my love to all of you,
Bud

Letter to Mom from Harland "Bud" Curtis

Camp Mackall, North Carolina
Friday, May 5, 1944
Headquarters, First Battalion

Dear Mom,

Gosh but it was wonderful to hear all your voices again last night. I am glad to hear that everything is going along fine and that Bert is feeling better. I was so anxious to talk to Jill, I forgot all about Lorrain. I hope she is alright. I am sure glad that Dad got home so that I could talk to him too. I hit that just right didn't I?

We will leave here tomorrow night for someplace, but it won't be far, because we arrive there Sunday and I guess it must be someplace in Virginia. My calculations are all shot again because it might be that there is such a wonderful girl as Jill to come home to, I can't help but come back.

I don't know whatever made me think Jill didn't love me. I'll never have any doubt about it ever again. Love her very much mom, she is the girl for me so take very good care of her for me and have her come out to the house as often as possible so you can write to me and tell me that she is alright and not falling off any garage roofs. Tell me about what she said after I hung up. How cute she looks and all that. Gosh, but I miss her and everyone, but I will be coming home again in just a little while, and all of this will just be a memory like all the other things like those long days in Toccoa, and that mud and rain in Tennessee. Wait and see time will take care of everything. Just write to me as often as you can, because I sure like to hear from you.

I hope you got the check I mailed? Be looking for a couple of small hand bags and packages as I have been sending home everything that is extra and in the way.

Well I had better sign off now. Thanks again for everything and for having Jill there last night. Take good care of her.

Lots of love,
Bud

Letter to Mom from Harland "Bud" Curtis

Camp Mackall, North Carolina
Friday, May 5, 1944, Headquarters, First Battalion

Dear Mom,

I just received this letter I am enclosing from Mrs. Doyle. I don't know what to think of it, because it all seems so fantastic that anything could happen to someone I know. I can see Dorothy's mother must think that there was something serious maybe between Dorothy and I. I like her a lot and if it weren't for Jill, I would probably like her much more, but no one can take Jill's place. Anyway, it kind of worries me to think such a thing could happen to such a nice girl as Dorothy.

I wish you would call up Mrs. Doyle and tell her I got her letter and that I am very sorry that Dorothy is sick. Don't say anything to her about Jill or anything like that, because I don't know exactly what Mrs. Doyle meant when she said in the letter. "She had always hoped that Dorothy and I would get together sometime, but is sorry now that it can never be." The way it sounds it makes it as though she thinks I was rather serious about Dorothy or something, but then again it just might happen to be the way she worded it and doesn't really mean it the way it sounds; so it is better if it isn't ever mentioned, O.K.

Just ask her how Dorothy is and tell her that I got her letter and that I was shipping out the same day so I didn't have time to write and tell her how sorry I am about it all, but *will write as soon as I get settled someplace. Write and tell me what she says and if "Dorothy is feeling better as if she still doesn't remember anyone and all that.*

I am going to have to hurry. I hope you get the other letter I wrote this morning and had two money orders and the check in it.

Write soon,   Love to all,   Bud

Letter to Mom from Harland "Bud" Curtis

Camp Mackall, North Carolina
Saturday, May 6, 1944, Headquarters, First Battalion

Dear Mom,

Well here it is Saturday and it will be the last day I will ever be in this camp again.

The latest rumor now is that we will leave Mackall tonight by boats and be part of the Navy. Well it is raining hard enough now that I wouldn't be a bit surprised to see the Queen Mary come floating up the main street any minute now.

I just went over to the post office and mailed another package and got a couple of money orders made out. I am enclosing a check from that train ticket I didn't use.

I hope they all get to you. Be sure and let me know when they do. I bought $2.00 of air mail stamps so I guess you won't have to bother about sending me some for a while.

Write as often as you can. Tell everyone hello and remember to take good care of that scatter brained blonde of mine for me.   Love Bud

Chapter 11
POINT OF EMBARKATION (P.O.E.)

Colonel Louis A. Walsh, Jr.
Major General, USA, Retired

Colonel Walsh, First Regimental Commander

Colonel Rupert D. Graves

Colonel Graves assumes command and takes the 517th PRCT to Combat

Just prior to deploying overseas Colonel Walsh was relieved of command of the 517th. It was a real shock to the men of the 517th. He had put all he had into making the regiment a first class fighting force. Unknown to most, he had frequently gone to bat with higher headquarters for his officers and men. On at least one occasion he had offered to quit the 517th unless charges he felt unjust were dropped. In speaking with Colonel Boyle at the Washington, D.C. reunion on June 30, 2007, he said, "Colonel Walsh would tell higher headquarters to Fart, fume, and fall back." At this reunion Major Don Fraser stated, "Lou Walsh was a good commander, I liked him a lot."

In 2005, Colonel Walsh's son, Trooper Walsh wrote on the 517th PRCT email site the following about his father,

"Some of you Soldiers will remember my Father, Lou Walsh - the Commanding Colonel who activated and molded the 517th Parachute Infantry Regiment at Camp Toccoa, Georgia, starting on 15 April 1943.

History was not always kind to Dad...years later he joked that he was the first and youngest Full Colonel in his class at West Point, and he was also the first in his class to be bumped down a notch from Full Bird...all this revolving around an issue where Dad one day stood up for his troops to a General...a fella who was a class ahead of Dad at the Point, and not one of his fans or strong supporters (throughout life people either loved or hated Dad - sometimes both at once).

Those of you who were there at the time can feel free to correct any of my details, but as I heard and remember the story it goes that the 517th had just come off two day maneuvers. The men were all tired and dirty when this General showed up and ordered my Dad to prepare for a white glove inspection. Dad, using his... ummmmm - unparalleled less than quiet and calm demeanor, promptly instructed his superior officer as to which orifice he could stick his head up for a better view - or something to the effect. Not surprising, Dad was relieved of command and this shortly before the 517th was to leave for Europe, under the supervision of Colonel Rupert Graves. Now to every action there is an opposite (and sometimes positive) reaction. In my Dad's case the positive reaction was the immediate expression of undivided love and respect from his men which was to help carry Dad throughout his life over the next fifty-some years. Some of those friendships proved to be active until the day of Dad's passing. In particular I think of some of the fellow officers my Dad cherry picked to form up the 517th...like with Dick Seitz and John Lissner. There is no other outfit like the 517th Parachute Infantry Regiment." - Louis "Trooper" Walsh, son of Major General Louis A. Walsh

Trooper Walsh and Bud Curtis at the 517th Reunion, Washington, D.C., July 1, 2007

The new Regimental commander had some big shoes to fill. Lieutenant Colonel (LTC) Rupert Graves would take the outfit overseas. Reasons for Colonel Walsh being relieved were never disclosed and LTC Graves did not discuss the subject with anyone. LTC Graves, a older man who was in his early 40's graduated from the United States Military Academy (USMA) class of 1924, and came from the 551st Parachute Infantry Regiment. Archer (1985) said, "LTC Graves was cool, calm, unassuming, and professional. Like all new commanders he was closely watched by the troops, treading carefully and with great circumspection. LTC Graves quickly proved to be likeable and unflappable, and before long things settle back into normalcy." LTC Graves was promoted to full Colonel. He did remain the commander from the time they deployed overseas until the Regiment was deactivated in 1945. Colonel Graves had this to say about his Regiment.

"The members of this combat team were young Americans of that period where even the most senior battalion commanders and regimental staff officers were young graduates of the Reserve Officer Training Corps (ROTC). What motivated these people to close in against well-defended enemy machine gun positions, often manned by elite Nazi SS troops, may be open to question. Perhaps initially it might have been patriotism, but later this was replaced by comradeship acting as a sort of life force inducing a loyalty that transcends friendship. Simply, they did not want to let their pals down. Good physical conditioning from long, hard training, and the spirit of optimistic youth willing to take a chance was also another probable ingredient. Although I have served with many different military organizations during my years of service, I have never served with any group that compares with the spirit, the initiative, the physical qualifications, and comradeship of the 517th PRCT. I feel grateful for the opportunity given to me to take this unit overseas and bring it home" (Paratrooper Odyssey, 1985).

Col. Louis A. Walsh, Jr. (Maj. Gen. USA Ret.)   Colonel Rupert D. Graves

Colonels Louis A. Walsh first Regimental Commander, and Rupert D. Graves who took the unit overseas and brought them back to the United States

In early May 1944, the 517th staged at Camp Patrick Henry, Newport News, Virginia. The troopers milled about more or less for ten days, making their acquaintances with some lovely members of the WAC contingency. Bud writes home about meeting these WAC on the ship over to Italy. On May 17, 1944, troopers of the 517th boarded the gangplank for their great adventure on the ship Santa Rosa. A Chief Warrant Officer Everett Schofield, Regimental Personnel Officer was at the rail with a few other troopers. Chief Schofield had served in World War I, twenty-six years earlier and was heard to say, "Take a good look. This outfit will never be the same again."

Letter to Mom from Harland "Bud" Curtis

Somewhere on the East Coast
Monday, May 8, 1944
Headquarters, First Battalion

Dear Mom,

I am at a camp on the east coast. I don't know how long we will be here, but it would be O.K. with me if we stayed right here for the duration, because this is sure a pretty place. Big green oak trees all around and shady just like a big park. That cool ocean breeze sure reminds me of home.

Did you get my letter I sent from Camp Mackall? One had a check for $120.00 and the other had a check for $26.40 and two money orders; one for $25.00 and the other for $100.00. Let me know if you get them. I also mailed a few packages of excess junk I had hanging around that I didn't have room for.

I guess from now on I will have to be asking more questions about whats going on around there, because I can't tell you very much about what I am doing because of military secrecy, so if my letters are short and uninteresting I know you will understand. Just don't worry about me because I can take care of myself and even if I couldn't I am in the best outfit in the world and with the best guys in any man's army, so there isn't any need to ever worry about me. I'll have a lot of interesting experiences that I could never get in any other way, but I will just save them up and tell you all about everything when its all over and I come home again.

I like this First Battalion much better than Regimental Headquarters that I was in. The fellows all seem to be pretty good eggs and from what I have seen of the other companies, we have the best first sergeant and company commander of all of them, so I think I should get along fine.

I should be getting mail from you pretty soon now. Send all your letters to the above address from now on.

Have you taken any more pictures of Jill or anyone? I hope so.

Well I will sign off for now and will write again soon.

Lots of love

Bud

P.S. Better send all your letters airmail (its much faster)

Letter to Mom from Harland "Bud" Curtis

Somewhere on the East Coast
Thursday, May 11, 1944
Headquarters, First Battalion

Dear Mom,

At last the mail is beginning to catch up with me, and I got two letters from you today. It was sure good to hear from you and find out what you have all been doing. I am sure glad that I did get a chance to talk to everybody before I left. I can call up from here but there would be someone listening in all the time waiting for me to make a slip and say something about where I am at, and then I would be cut off. I am hoping that we might get passes out of here and then I would call up. I still couldn't say anything about where I am, but I would feel much better about talking, knowing that there wasn't one of these southern half wits listening in on the conversation. If I don't get a pass, I will try calling from here anyway. About all I can tell you is that I am on the east coast, but not in New York.

So they got Don Donahue *(Bud's Uncle Emery and Aunt Cassie's son)* stuck out on New Caledonia now. I can't say that I would like to be put on any island out in the middle of no where like that. Garth is also stationed on New Caledonia. He has been there for quite a while now. Why don't you give me his address and I will send it to Garth, and he can look him up. Better yet, I will give you Garth's address and Don can look him up. It is Garth Boyce S 2/c GM., Air Center, Navy 145, C/O Fleet Post Office, San Francisco, Calif. I am pretty sure they know each other.

Why don't you give Jill a book of air mail stamps as a present, so that she will be sure to send all her letters airmail. It takes an awfully long time to get mail that is just regular, and the farther I go, the longer it is going to take, so remember to send them airmail from now on. Thanks. Don't send any V-mail letters until I let you know for sure that I am not in this country anymore. Sometimes airmail is faster anyway.

I am glad to hear Bert has finally got his motorcycle put together and is feeling better himself.

I hope that you can use the money I sent home. I am sure that Bert could use it. I owe you all so much anyway, but I guess I will eventually get it all paid back I guess it has been rather hard for you to have Jill come out to the house with so many things happening, but have her come out as often as you can and then write and tell me about it.

I guess I will finish this up now and I hope I get another letter soon.

Lots of love

Bud

Letter to Mom from Harland "Bud" Curtis

Somewhere on the east coast
Mother's Day, May 14, 1944
Headquarters, First Battalion

Dear Mom,

I have been trying to get a call through to you all evening, but it is now 2:30 am and there is still over 4-hour delay, so I guess I won't be able to talk to you. Everybody in the services must be calling home because it is Mother's Day, but there will be a lot who won't be so lucky like me. I sure wish that I could talk to all of you again, because I won't be able to write for a while after tomorrow. Military reasons and stuff, but as soon as I can write again I will let you know as much as I am allowed to. Please don't worry about me well you! There are a lot of things that I would like to say, but I can't. I guess I explained most everything in my letters I wrote at Camp Mackall though. Maybe next Mother's day I will be home again. Let's hope for the best anyway. I am not so good at saying a lot of pretty things and stuff, but you are the best mother in the world and I love you very much. It is much easier to feel it inside of me than to write it on paper. I think you understand. No matter where I go, what I do, or how long it might be that you don't hear from me, just always remember how much I love you and it won't be very long until I will be home again and in a couple of days you will forget I was ever gone.

I am awfully sleepy, so I don't think I will even attempt to write Jill a letter. Just tell her I love her and not to worry if I don't write. Take good care of her for me and tell her the next time I come home I might have had a part in making this a better world for that Geri junior to be celebrating Mother's Day in.

I would hate to think that any of our off springs will ever have to see any part of a war so I will be doing my best to see that things will be different by then. Enough of this talk. I feel like I am waving a flag or something, so I will say so long for a while.

Lots of love to all of you,

Your son Bud

517TH PARACHUTE INFANTRY

Camp Toccoa, Ga.
6 May 1943.

Dearest Mother:

On this happy day, I want to thank you sincerely for all you have done for me. I feel helpless to repay you. You labored and toiled for me night and day. You worked hard to make my life happy. Everything that's best in life--I owe to you. I shall never forget everything you did for me.

Love Bud H

MOTHER OF MINE

God's fingers painted the dawning
 And traced all the silver there,
But dearer to me is the silver
 He laid on my Mother's hair.

Ah sweet is the light on the waters
And blue are the summer skies;
But sweeter and deeper the love-light
 That shines in my Mother's eyes.

The sun sparkles bright on the dew-drops,
 That lie in the rose's vase,
But Heaven's own beauty awakens
 In the smile on my Mother's face.

Oh God, the works of thy fingers
 Are wondrous and so Divine;
But this is the earth's rarest beauty--

 That wonderful MOTHER OF MINE!

Rev. Robert Kearn, C.SS.R.
(Chaplain overseas)

The 517th PRCT Chaplain Robert Kearn prepared a poem for Mother's Day and gave it to all of the troopers. Bud signed it and sent it to his mother.

Chapter 12
Overseas Combat Duty

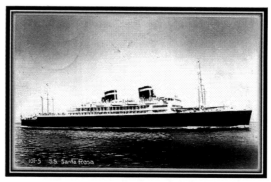

The S.S. Santa Rosa took the 517th PRCT from Newport News
Virginia to Naples, Italy, docking on May 31, 1944.

As the Santa Rosa sailed down the James River the troop ship was joined by a Navy destroyer and the convoy headed out into the open Atlantic. The odyssey of the 517th had begun. As the ship crossed the Atlantic, rumors were they were heading to England to prepare for the invasion of Europe. However, on a dark night the ship slipped through the Straits of Gibraltar and it became obvious that the destination was Italy. The Santa Rosa docked at Naples on May 31, 1944. Troopers filed down the gangplank and were loaded on to railroad cars and were taken to Neapolitan, a suburb of Bagnoli. Enroute Colonel Graves was handed orders directing the 517th to attack from Valmontone to Rome the next Day. None of their crew served weapons were available. Artillery and vehicles were not there yet, and the attack would have to be with rifles only. When this was pointed out to the higher command the mission was cancelled. The 517th moved to the "Crater". The "Crater" was a bed of long extinct volcano that was rumored to have been the private hunting reserve of Neapolitan royalty. The order was given to set up tents, and wait for the crew served weapons and vehicles to arrive. It would be two weeks before they arrived. Bud describes in a letter to his mother about swimming in the waters of the "Crater"

Letter to Mom from Harland "Bud" Curtis

Somewhere in Italy
Wednesday May 31, 1944
Headquarters, First Battalion

Dear Mom,

Well here I am, I'll bet you have been wondering why you haven't heard from me for so long. I took a little jaunt over to this side of the world to see how the other half lives, and all I have to say is I sure wish you folks at home could see me now - - "At Home" get what I mean? The weather is nice and warm anyway.

There is something that has been worrying me. Remember when I was home last furlough? Well when I sold my car to Jim I was going to get a receipt for the darn thing and never got around again to see him. Just so there won't ever be any legal details to arise about it, I wish you would drop around to his house sometime and get the receipt for it. I have lost his address, but he lives on Corinthian Walk, the second house on the east side of the street at the beginning of where it comes to a dead end. You can't miss it. It is right between Belmont Shore and Seal Beach! Let me know if you get the receipt and if you find his place okay. Thanks!

Boy am I having a heck of a time writing with this Red Cross pen, but try to make it out and next time I'll do better. This seems to be a pretty good place as any I guess, but I don't know how long I am going to be here. Lots of rumors but nothing definite yet. Boy, what I wouldn't give for a glass of milk. Some of the guys around here say they haven't seen *any milk for 15 months. This place reminds me of the good ole year of 33 (1933 was the year the earthquake struck in Long Beach. Bud and his family lived out on their front lawn for two weeks before things got back to normal), but I guess I can live through it.*

I guess I have covered most everything so I will close for now and see what is for chow. Don't worry about me if you don't hear from me so often. Things are just that way. Keep your fingers crossed and hope for the best because I think I will be needing it. Tell Jill I love her and all my love to all of you.

Bud

In April 2005, Bill Houston, a former private first class (PFC) with the 517th wrote in the periodical, Thunderbolt, first quarter 2005, page 39 " It was on the thirty first of May that we docked in Naples, Italy. We loaded into railroad cars that took us to Bagnoli, a suburb of Rome. From the railroad station we hiked to the Crater, an extinct volcano which it was claimed that it was hunting grounds for Italian royalty. The crater was a picture postcard, round full of lush vegetation and about 800 feet below the rim of the volcano. On the floor of this beautiful crater the 517th set up their pup tents all in neat straight rows. A road ran along the steep slope of the crater and up to the top. Sometime during the first or second week of June, we received our first payday, and it was paid to us in military currency. Money printed by the Allies for use in occupied countries. When we went to town the Italians would not take Italian currency. They wanted the military currency, but when it came time to give us change they always tried to pass the Italian stuff (change/currency) off on us. At first we did not know what they were trying to do to us, but soon we learned, and from then on we refused to accept the Italian stuff."

V-Mail Letter to Mom from Harland "Bud" Curtis

Somewhere in Italy
Saturday June 3, 1944

Dear Mom,

How is everything over there? Nice warm weather by this time I'll bet. I have been on guard duty so I haven't been around our bivouac area for a day now, and when I go back this afternoon I should have a letter from you – I hope. I have gotten one that was addressed to my APO (Army Post Office) number.

Has all of that stuff I sent from MacKall got there yet? Chances are I might run into Lieutenant Willard Hill over here. (Willard Hill was the Paratrooper Bud saw at Church and impressed him so much he joined the paratroopers). I have met several guys that we in his outfit and know him, but they don't know where he is right now. I will let you know if I see him. I will close for now and write a long letter soon. I have written Jill several times so between the two of you – you can keep up pretty well on "what's cooking." Bye now. Lots of love - Bud

V-Mail Letter to Mom from Harland "Bud" Curtis

Somewhere in Italy
Tuesday June 6, 1944

Dear Mom,

Still no mail since that first day, but maybe I will get some today. Things are still about the same. Not much doing right now. You can read in the paper and find out more what's going on round here than I could tell you.

It is really awful the way people live here. You know how dirty the streets are in Mexico; well it is about the same way here, only things are a bit war torn and most of the buildings look like the remains of the way the buildings did after the earthquake we had in Long Beach.

That kind of gives you an idea what a place looks like after it has been bombed. I will take my camera into town the next pass I get and get some pictures of non-military ruins and stuff. About that deal of Dorothy. I had an idea that was about what the setup was, so forget all about it now. Take good care of yourself, and I will be home again as soon as we put Hitler where he belongs. Bye now, Love Bud

In the above letter dated June 6, 1944, my father, Bud Curtis referenced a girl named Dorothy which he tells his mother to forget about. It does not seem to make sense until you know the rest of the story, and I would have never know the rest of the story if it was not for Rick Sweet. Rick's father was Odas Sweet who served in "H" Company of the 517th PRCT. Because of our common bond of having fathers in the same parachute regiment we became good friends. On Friday, December 14, 2007, at approximately 8:30 a.m. Rick telephoned me. He said, "Did you know a 517th Christmas card with sterling sliver paratrooper wings attached to the card is from your Dad addressed to Dorothy Dole, and is being auctioned on Ebay"? I said, "No I do not even know of a Dorothy Dole, let me call my Dad and ask him and then I will call you back." I then telephoned Dad and he told me the following story.

"In about 1940 or so before the Pearl Harbor attack, my brother Bert asked our parents if he could take the family car for a drive. They said okay, and Bert got one of his friends to go driving with him. I asked Bert if I could go with them. He reluctantly said yes so I got in the back seat and they drove from our home at 3714 East 6th in Long Beach down by the Lagoon on Colorado Street, approximately a mile or more away from our home. The Lagoon was built to host the Olympics in the 1930's for diving and swimming events. It was now a popular swimming place for local citizens.

Somewhere past the Lagoon Bert stopped the car, then asked me to get out and check to see if the left rear tire was flat. I got out of the car and went to check the tire. Bert quickly stepped on the gas pedal and sped off with his friend laughing. They never came back to get me so I started walking home. Just then a big large black car (limousine) like a Packard stopped, and the man driving asked me if I needed a ride. I did not know it at the time, but the man was Clyde Dole, a United States Congressman from California. I quickly said yes, and got in the backseat where the Congressman's daughter Dorothy Dole was sitting. They drove me up to Grand Avenue and 4th Street a couple of blocks from my home. Although I do not remember now, but I must have obtained Dorothy's address and phone number.

Sometime later I called Dorothy and asked if I could come visit her. She agreed and I went to her home. I had never seen such a large home with lots of expensive things. Dorothy's folks were real strict and would not let her go out on a date, but allowed Dorothy and I to visit together in her home. I returned to see her on several occasions.

In April 1943, I left for paratrooper training at Camp Toccoa, Georgia, and later was transferred up to Camp MacKall, North Carolina. In October 1943 I came home on furlough to California. I again visited Dorothy and her father let me take Dorothy in his family car. I drove out to the Army Airfield in Santa Ana to make arrangements to fly back to Georgia on a military aircraft. I was very surprised Dorothy's father let me take his car and Dorothy on a short day time date.

My furlough was soon over and I returned to Camp MacKall, N.C. I continued to write Dorothy letters. The holidays of Thanksgiving and Christmas were fast approaching and I had nowhere to go for the holidays. I spent most of the time by myself in camp while the troopers from the East and Midwest went home for the holidays. Although I do not remember now I probably sent the Christmas card to her during sometime in December 1943, before Christmas.

Shortly, thereafter Dorothy's mother wrote me a letter telling me Dorothy was very sick. She said I should not write Dorothy any longer because of her illness. I figured it was because she was such a high society person, and Dorothy's parents didn't want her being involved with an Army Paratrooper who had no credentials. I honored Dorothy's mother's wishes and never wrote here again.

After the war in 1945, I was shopping in a store in downtown Long Beach. There I accidentally encountered Dorothy who was also shopping. We visited briefly where I found out she was now married. After that meeting I never saw her again."

Now sixty-four years later the Christmas card with his very expensive sterling silver paratrooper wings attached were being sold on Ebay. It was hard to believe Dorothy would have saved this card and wings all of these years. I quickly told my family about the card and wings. One of my sons emailed the person selling the item on Ebay and told him, "I am the grandson of this paratrooper, name your price and I will pay it." No reply. Then on Sunday, December 16, 2007, my father passed away very quickly. This story was the last thing he and I ever discussed. How ironically the last conversation I would ever have with my Dad was about the 517th PRCT. Shortly thereafter my daughter-in-law emailed the seller of the card and wings, stating the paratrooper had just died, and the family really wanted the card and wings. Once again he refused to answer our email and on the last day of the auction a person paid $300 for the Christmas card and sterling sliver paratrooper wings. Never in my life did I ever think something of my Dad's would be auctioned on Ebay. I was amazed that Dorothy would have saved that card and his paratrooper wings all of these years. On page 78 you can see the same 517th Christmas card that Bud sent home to his mother. The card was an expandable card showing the different phases of training the 517th went through. Embossed at the top are paratrooper wings. This is where Bud attached his wings to Dorothy.

Christmas Card sent by Bud to Dorothy Dole December 1943

Letter to Mom from Harland "Bud" Curtis

Italy
Saturday June 10, 1944

Dear Mom,

I just got your letter that you wrote on May 30th. That's not so bad only 10 days to get here. But I think there are a few letters that I haven't received yet, because in this letter you told me about Bert being out on some island after taking another cruise. I still thought he was working in the ship yards. The last letter I got you said he was lead man out at Cal-Ship. So I was rather surprised that he shipped out again. If he can work out in the ship yards and have a deferment; why the heck doesn't he stay there? It doesn't matter particularly where I go and there won't be much loss if something happens to me, but I don't like the idea of Bert taking chances, and especially when he is going to be a proud father pretty soon. As long as he can keep deferred from the draft he will be a lot better off staying right there in Long Beach. I think he should stick close around home now more than ever. If it ever comes to a point of the Army or the Merchant Marines, what ever he does, have him stay out of the Army.

It's O.K. for a guy like me, because I am still single and haven't any responsibilities and Bert has. If it's the money part of it, well I will send as much home every month as I can and he can have all of it. Things are kind of hot over here as you can find out from the newspaper and it won't be very long until I am right in the middle of it. The sooner the better. Anyway my chances are just as good as the other guys of coming out of it, but this that goes visa versa, so hang onto all those papers just in case something does happen to me and you can collect a $10,000 insurance out of the deal, but don't worry about that. I sure hope that I get back, but the part of maybe I won't doesn't worry me in the least if my time is up, there is nothing anyone can do to stop it. I just hope that if something like that does happen that I will at least have had a chance to take few Jerry's along with me to keep me company when I am down there shoveling coal.

I came in this outfit to fight and it has taken a long time for us to get over here and now that I am here and have seen with my own eyes what is going on, and talked to guys who have been here for about 15 months and the guys who have been up on the front lines, well, it makes me want to get in and get it over with, just that much more. So if anything ever should happen to me while I am in there doing my best to end all this so people can go back to living like they want to and be civilized again. Don't ever feel bad about it. I wouldn't want so much as one tear shed over me. Instead I feel proud that I could do my part and think of me just as I was adventuring around looking for excitement. I like to travel and I imagine there is quite a bit of it in between this world and wherever you go from here. So always think of me as being out traveling around and looking for some new places to go. There I am glad I got that off my chest and now that you know how I feel about that part of things, I am sure you feel better about it too. So now forget about it, because I have all the intentions and determination there is to get back to good ole Long Beach again.

How did that shower for Lorrain turn out? I am glad to hear that maybe Jill won't maybe be going up to Santa Barbara to college. I would much rather she stayed right there and then she can come out to the house every so often. I sure wish I could be there when she graduates.

I was thinking of getting a kind of decoration table cloth that I saw in town the other day and send it to Jill to put in her hope chest, but you said in this letter that you had got a new table cloth and I am afraid you would feel kind of slighted if I sent it to her, and I don't have quite enough money for two of them. I'll figure out something. Borrow some money or something. Don't send any until I ask for it because it would take too long for it to get here and I might be someplace else by then.

I'll bet the place really does look good after you finished decorating it. I'll be around to see it before long maybe.

I haven't got the letter that Jill sent with the pictures yet. But I will probably get it soon. I am holding out for that big one that she will have taken in her formal. I wonder who the lucky guy will be who takes her to that prom?

Did Bert take his motorcycle along with him this time. What rating is he now. Still an Ensign? Maybe there is a letter I haven't got yet that has all those answers. How is the mail situation there, are you getting my letters O.K.?

Every two weeks we get rations, 6 candy bars, one pack of gum, a coke, if they have them, one pack of razor blades, and a bar of soap. It would be nice if you would send some of these cookies and candy, but I am afraid that they wouldn't be much good by the time they got *here. Did you get all the things I sent from Camp Mackall? The Mother's Day card and all those boxes and stuff? Did Jill get that box I sent her?*

If Bert ever gets over this way have him look me up. He would just have to ask around where the 517th was. News travels fast around here. I have met several fellows that know Lt. Willard Hill (Willard is the one Bud saw at church down at Ximeno in 1943 in paratrooper uniform and it made dad want to join the paratroopers), *but they say he is in England now. Probably he and Dean Hildreth* (friend from high school) *were in on the invasion. Wish I could have been.*

Well, write soon and tell me what's the latest.

All my love

Bud

The unit was about to have their first combat experience. Colonel Tom Cross, U.S. Army, Retired, and then a member of the 517th PRCT recounts the organization of the unit. "The 517th consisted of three parachute infantry battalions: The 1st Battalion commanded by Lt. Col. William J. Boyle, the 2nd Battalion commanded by Lt. Col. Richard J Seitz and the 3rd Battalion commanded by Lt. Col. Melvin Zais. The 460th Parachute Artillery Battalion of the 517th PRCT was commanded by Lt. Col. Raymond L. Cato and the remaining contingent of the 517th PRCT, was the 596th Parachute Combat Engineer Company, commanded by Captain Robert W. Dalrymple. The 517th PRCT arrived from the States after just having completed its final large scale airborne training exercise and participation in the Tennessee Maneuvers." The 517th PRCT was specifically attached to the fifth Army for ten days of battle experience in the lines."

On June 14, 1944, the outfit struck tents and moved to the beaches of Naples, Italy and waited to board LST's to carry them to Anzio. At noon the LST flotilla pulled into shore. The Navy crews quickly left their boats, went into the Alligator Club, and were not seen again until dark. They came back and the 517th loaded onto the LST's. They sailed northward, but their destination was changed. They were to land at Anzio, but the Germans were retreating too fast. They now were on to Civitavecchia, and landed there the next day about noon. The troops marched off and set up a bivouac site several miles inland. They didn't stay long and were quickly assigned to the 36th Infantry Division. This division was from the Texas National Guard and had performed well in combat. On the 17th of June, the 517th was brought to the south of Grosseto. Troopers dug foxholes, and prepared to stay, but they didn't, they were now marching to Grosseto. The Germans had now given up Grosseto, but now they were north of the town on the Ombrone ridgeline called "Moscona Hills." At day break on June 18, 1944, rifle battalions moved out to attack Moscona Hills.

The order of March was First, Regt HQ, Second, and Third Battalions. The lead battalion (First), of which Bud served in ran into a storm of machine gun fire as it entered the Moscona Hills. Troopers fanned out, took cover, and returned fire. The first casualty of the 517th was Sergeant Andrew Murphy of "B" Company, He was killed by a sniper. The First Battalion was pinned down and could not move. The Second Battalion came to their rescue, and with artillery drove the Germans out. Once the battle was over the men of the 517th were startled to find out that the men they thought were elite blond hair, blue eyed Germans were actually scruffy undersized Mongolian type Muslim minorities from the Soviet Union. The Germans had captured these Mongolians, and forced them into fighting for the Germans as part of the 162nd Turcoman Division. Bud discovered German soldiers a few yards away were observing the Mongolian soldiers, and if they did not fight the Americans they got a bullet in the back by a German solider. The first day of combat resulted in 40 to 50 casualties. Bud recalled his first experience in combat. He said, "I was standing over by a truck and heard what I thought were bees flying around my head. It sounded like a crack and then another, and another." He then heard what he now knew were bullets flying around his head and hitting the side of the metal of the truck. He quickly fell to the ground and took cover. Bud recounts that most of the troopers had no idea they were being shot at. They ran bravely from position to position advancing toward the enemy. After the battle was over the German commander told his captors that he had never seen such bravery as these paratroopers would not take cover, and were not afraid of being shot by being continually exposed to enemy fire. Fortunately, the Germans never knew these were green troops experiencing their first combat and bravery was not their motive. It was ignorance of the dangers that existed. After the battle, Bud had his picture taken over a dead enemy soldier. It was the Mongolian sniper that had killed Sergeant Murphy. Bud posed as if he were a big game hunter over the dead Mongolian. Toughening up emotions by these sorts of antics would help Bud, and the other men from thinking about their part in having to kill another human. In a soldier's mind they were not human but less than human, they were enemy soldiers who were trying to kill them.

On June 19th the Second Battalion captured the hilltop village of Montesario, and the area was pretty much secured. The 517th moved onto a ridgeline south of Gavarrano and bivouacked overnight on June 22nd. During the night of June 24th the Third Battalion made a long infiltration, emerging next morning on high ground overlooking the dry streambed of the Cornia River. That morning at 8:00 am Bud and the First Battalion moved through the Third Battalion's area to seize Monte Peloso, dominating a broad valley overlooking the town of Suvereto about a mile north. The First Battalion continued the attack moving in column along a dry streambed. Under the cover of smoke the battalion moved to silence enemy positions. The fighting was very fierce because the enemy force was the elite 29th Panzer Grenadier Division. The best the German Army had. None of these Germans would surrender. Enemy artillery fire continued on Mote Peloso through the night. This would be Bud's first experience with being on the receiving end of artillery fire. It would not be the last, and would affect him for the rest of his life. In the morning the 517th was to attack Suvereto, and the word came down to the men to fix bayonets. Bud remembered hearing that command was unnerving. He felt this could be his last night on earth. The German Panzers were tough, and he knew it would be a bloody fight. Then at 0400 hours (4:00 am) a phone message came over the wire "Daddy is here." What did this mean? No one in the command had any idea, but it became clear when the Third Battalion of the 442nd (Nisei) Regimental Combat Team showed up to take over fight. These soldiers were from Japanese American families who were taken from their homes after the attack on Pearl Harbor, and put into camps in many different states in the west. The young Japanese Americans became known as Nisei and quickly volunteered to fight in Europe. The 442nd Regimental Combat Team was formed. The attack took place in the morning, but the 517th was relieved by the 442nd who moved up to close in on the Germans. They took heavy causalities. Bud remembered seeing these wounded men being evacuated to the rear. He was very grateful he did not have to fight that battle. He gained great respect for these tenacious warriors of the 442nd. They undoubtedly saved Bud's life.

Captain Charles La Chaussee, Commander of C Company, wrote about this first battle. He said, "Soon after dark the enemy began shelling the hill with artillery from beyond Suvereto. The shelling continued sporadically until dawn. At daylight, khaki figures began to move up that hillside from the rear. They were short rugged looking types, and to our surprise were Japanese. This was the 442nd Nisei Japanese American Combat Team in its first commitment to action. When the 442nd was in place we filtered back off the hillside, formed in column, and moved to the rear. Although we were unaware of it at the time, this was our last action in Italy." Thunderbolt, first quarter, 2005, page 41.

V-Mail Letter to Mom from Harland "Bud" Curtis

Deep in a Foxhole - Somewhere in Italy
Friday June 22, 1944

Dear Mom

I won't have time to write but this one letter as it will have to be to all of you. I got your June 7th (1944) letter today. Jill's too, and was I ever glad to hear from you. It seems that you thought I was in on that D-Day in France (Bud is referring to June 6, 1944, the invasion of Normandy). No, I wasn't there, but I am in the middle of a D-Day here that is just as bad. I have seen plenty of action in the last few days and I thank God that I am yet still alive, and I pray with everything in me that I will be home again. So far in my life I have never experienced anything as terrifying as having artillery shells dropping all around me. A sniper bullet whizzing past my ear is mild to the screaming of an artillery shell and the jagged shrapnel that is death to anyone in its way. I hit the ground and flatten out like a pancake and pray with all my might that I don't get hit. It is hell, but we are doing a good job and no matter how scared I get it I always know that you all at home are pulling for me and have your faith and trust in me that whatever comes, that I will meet it bravely like the man you have raised me to be. With that thought I remain as cool headed as I can and trust in God. I have never before been so serious about things in my life. Life is cheap around here and if I do get back I am going to be most attentive church member there is. This may be my last letter to you. God only knows, but whatever happens just remember I am doing my best to end all of this. Tell Jill I love her with all of my heart and that she will never know how much that means to me. It is Jill alone that drives me on when everything inside of me feels empty and dead. My only thoughts now are kill or be killed, and to get back home, and forget all the terrible things I have seen. I won't be able to write you very soon. It is hard to tell you not to worry when I can't convince myself, but remember I love you all and your letters mean so much to me now so keep them coming. Keep that Icebox full because when I get home I am going to eat for a week straight and then sleep the whole next week. I'll give em hell Dad. "Love from a Paratrooper." Bud

A V-mail letter was a letter written by the solider then photograph and reduced to a size of about 4X4 inches, then sent by the War Department to the addressee for free.

Letter to Mom from Harland "Bud" Curtis

Italy
Wednesday, July 5, 1944

Dear Mom,

In a few days I will be able to tell you where I am at now but all I can say for now is that I have had better living conditions these last few days than I've had for a long time. Right now I am listening to a radio. It is even playing good ole American music. I actually heard a rebroadcast of Bob Hope last night from the Naval base there in Long Beach. But if you think that's something just hold on and I'll tell you more.

There was a U.S.O. show that played for us today and there was a girl that played the accordion and she came from Long Beach. She was a couple of years older than I am. Her name is Barbara O'Connell. She was a senior at Poly (High School) when I first started school there as a sophomore. She says she thinks she might know Bert from a physics class from J. C. I talked to her for quite awhile after their show was over. She knew lots of people I did and was even taking accordion lessons from the Verdugo Studios at Humphreys the same time I was and knows all those people there. She said that her folks had moved to Oregon, otherwise I would have given her our address at Long Beach and had her look you up and tell you she saw me over here, but she will be going to Oregon instead of California. That's the first American girl I have talked to since I left the States, except for those WAC's (Women's

Army Corps) that came over on the same boat with us. It was really nice to talk to an American girl and especially one that was from Long Beach.

    I had chicken for supper tonight. I even took a warm shower yesterday too. I will write soon and tell you more later that I can't now so I will close with all my love.

    Bud

P.S. Did you ever get that mother's day package and card I sent from Camp MacKall? Jill says she didn't get her package and I had a lot of expensive stuff in it that I bought for her. I don't remember if I had them insured or not. So if you didn't receive them I guess they are lost for good. I hope not though because I had a lot of nice things in both of them. You and Jill's and it all added up to about $50.00 I guess. Thanks a million for getting those nice things and orchid for Jill for me. It makes me feel much better. I can hardly wait for the pictures of her. Especially the colored ones. How about that big one I asked for. Have I got much of a chance of getting one? Well I guess I had better close for good this time.

    Love again

    Bud

    The 517th was initially sent to Italy to be part of an Airborne attack with the Seventh Army in operation ANVIL. This operation was to occur at the same time as the D-Day invasions of the Normandy beaches, called operation OVERLORD. One small problem had developed. Not enough equipment or airplanes to conduct two operations. Operation ANVIL had to be postponed. General Eisenhower now did not feel that Italy or the Mediterranean were major objectives in fighting the Germans. He felt a mass effort needed to be directed toward fighting in Northern Europe as soon as possible. General Eisenhower directed that Operation ANVIL be executed on August 15th. The British finally conceded and on July 2nd the combined chiefs of staff issued an order to go ahead with ANVIL renamed Operation DRAGOON. An order came relieving the 517th from IV Corps and directed them to be attached the "First Airborne Task Force" in the Rome area. Bud would start preparing for a combat jump into southern France near Le Muy, France.

V- Mail Letter to Mom from Harland "Bud" Curtis

Somewhere in Italy
Saturday, July 15, 1944

Dear Mom,

    I am just going to write a line or two just to let you know I am alright. I have been away from the 517th for over 2 weeks now, and I won't get any mail until I catch up with them again, so I don't know much about what is going on over your way. I should have a stack of letters waiting for me when I get back to the outfit and then I will write a long letter. I hardly know what to write about because it has been over 3 weeks since I have had any mail from you, Jill, or anyone. I have been getting to feel pretty good and have been getting a lot of rest. I'll write again in a couple of days when I get back to the outfit and get some mail and find out what everyone is doing at home. Tell Jill hello and that I will write to her soon.

    Love,

    Bud

Letter to Mom from Harland "Bud" Curtis

Somewhere in Italy
Tuesday, July 18, 1944

Dear Mom,

I am back with my outfit again and right now I am sitting in a cozy tent operating the switchboard. I guess Jill told you I wrote to her and said that I have been to Naples and went sight seeing at Pompeii. It was very interesting. We weren't situated so good then so I didn't get any souvenirs to send home. We are going to be here for a while I think and if I run across something I will try and send it home. I am getting mail regular now also. I have seen Rome and it is much better than Naples. Naples is very dirty and about ruined from all the bombings it has taken, but Rome is a very clean looking city and looks better than some cities in the U.S.A. Only the outskirts were bombed so the city itself looked swell. Close by is an enormous (big) volcano that is now filled with crystal clear water and is it ever nice to swim in.

Those packages you sent have a 100% chance of reaching me here so I should be getting them pretty soon. Thanks millions. Have you heard from Bert lately? How is Lorrain getting along? She is sure a nice girl and I don't think Bert could have found a nicer sister in law for me anywhere. Give her my love and tell her I hope it is a "*Claudia*" (referring to her first birth. It was a boy named Harland, but they called him Harley).

I am glad to hear that Reid is okay and making it alright. How does Jill like her job now? I hope she stays well now and doesn't catch any more colds.

Richard Adams wrote me and I guess he is doing fine. It was good to hear from him again. I think maybe I should have tried the Air Corps, but then I guess I wouldn't be satisfied living such an easy life as those guys do.

I have got enough airmail stamps to last me for a while so don't send any for a little while or I'll have so many I won't have anytime to sleep from writing so much to use them up. I have been getting the press too. I am enclosing some receipts that I have been carrying around see if you have all these things. I can't find the receipts of those two other packages so I guess they are just lost. On this rail exchange receipt there are two things, a canvas bag and that leather one. I had them both there on furlough so you know what they look like. Did you get them?

How about my watch. Have you taken care of that with Irene? I got a letter from her and she is going on a two week vacation, but I imagine she will be back by the time you get this letter.

Thanks a lot for the pictures and the ones that are on the way.

I guess that's all for now.

All my love,

Bud

Letter to Mom from Harland "Bud" Curtis

Somewhere in Italy
Saturday July 29, 1944

Dear Mom,

I just received a small package from you dated June 20, 1944. As for the candy in it, it was swell and the trip over here didn't seem to have bothered it at all. Do you have to have a request from me to show the post office for just a small package like that or is it just for large ones? If so, I will ask for one everyday, because it sure tastes good. I hope I get one of the *large ones pretty soon as this one today really hit the spot and has my sweet tooth craving for more.*

Cigarettes over here are better than money as it seems these Italians can't get them. You can sell a package of cigarettes to them for 50 to 60 cents and when we get P.X. (Post Exchange) rations we can buy a whole carton of 10 packages for only 50 cents, so you can see the value of them is better than money. There is a lot of black marketing over here, so there is a bad penalty for getting caught at it, but my conscience doesn't bother me for giving a package of cigarettes to one of these women around here that do our laundry or to some person for a haircut. Money isn't much good over here as the people haven't many places to spend it themselves. Food and cigarettes are what they want. In bigger cities like Naples and Rome there are a few things where money comes in handy. I bought two pairs of silk stockings for Jill at $5.00 a piece. I don't know if they are going to fit or if they are the right shade, but I imagine she can easily find someone who can use them if she doesn't want'em.

Are silk and nylon still rationed there or can you get them now? I hope Jill gets them in time for her birthday and in the meantime I will look around and see if I can find something else she might like and a souvenir for you. Things like clothes and most anything costs quite a bit of money and I am trying to save all I can as I don't think this war over here can keep going very much longer and if, or rather, when I finally come home for good I want to have a little money saved up. How much have I got in the bank now? Of course whatever I send home and you can use it for something I sure want you to feel free to do it as I could never pay back all the money and things you have done for me in a million years. With the way the allies are closing in on Germany on all sides I am now beginning to think more serious on post war plans and frankly it has me a little worried about what I am going to do when I get home again (In less than a month Bud would make a combat parachute jump behind German lines into Southern France on August 15, 1944. The war would continue on for over a year). I don't know if I should go back to school or start getting settled into some kind of a steady job I want; except I want to get into something where I will be my own boss because when I get out of this deal, I never want anyone telling me what to do again. Even though I have been in the Army this long and doing my best to keep in line with military discipline I still have an independent feeling that I have brains enough to think for myself. Everyday I sweat out in the Army, makes me appreciate more the things that seemed so unimportant before.

I am sure going to be a home loving guy when this is all over anyway. I have Jill's picture sitting over in the corner. She sure is cute. Every time I look at her picture and think that the only reason I am way over here looking at a picture instead of being there looking at her is because of those darn Germans. Well I can't put into words what goes through my mind. It is hard to keep your morale up when there are so many things to tear it down. Sometimes I figure what is the use, and the future looks so far away it seems almost useless to go on, but deep down inside of me I know I have to keep going because what we are fighting for is far bigger than just me and my far off dreams so I continue on like a darn machine and hope for the best.

I have been getting your letters quite regular but I haven't heard from Jill in over a week. I know she is writing, but it is still disappointing day after day not hearing from her. I am sure glad I have her to think of and make plans for what I am going to do when I get home. I guess if it wasn't for her I would have gone nuts long ago. Sometimes I wonder if maybe I haven't already. The only thing I ask for is letters and pictures so send as many as you can.

Enough corn for tonight. I'll write again soon. It is hard to find anything to write about although I have a million things I could tell you, but I am not allowed to so they will just have to wait.

Lots of love and thanks for everything.

Bud

P.S. I won't be able to look up that Chaplin who you say is a good friend of Bill, because he is in a different part of Italy than I am. Maybe I'll run into him sometime and who knows, maybe I might even meet Bill over here too.

As Bud prepared for Operation DRAGOON he and the men of the 517th were bivouacked west of Frascati, Italy. During this time the 517th took a moment to remember the men who were lost in their first combat experiences. Losses had been relatively low during the brief periods of combat. 129 casualties had resulted so far with 117 men being killed. On July 9th a memorial service was held and Colonel Graves made a brief address. During this time the men were living in pup tents. To cook their food, boil water for shaving, and other needs, the men scavenged for wood to build fires. Each day when the fire was no longer needed it was stomped out so the wood could be saved for the next day. One day Bud took his turn stomping out the fire. As he stomped on the burning wood he did not see a large nail protruding from one board. As he stomped on it the nail penetrated his right foot. This was a large nail and just about went all the way through his foot. Bud quickly pulled the nail from his foot. It hurt terribly, but he would not tell anyone. Maybe he was embarrassed, but more so than that he would not report on sick call because he did not want to be taken off of the mission for the combat jump into southern France. For weeks he hobbled around on this sore foot with the nail hole still in his boot. When the day came to make the combat jump his foot still was hurting and his boot still bore the nail hole. He never did tell anyone about this accident. He would not be left out of this mission.

August would be a rough month for Bud. His unit was scheduled to make the invasion of southern France. In a Associate Press release in the newspapers nation wide on May 16, 2004, Mort Robenblum reported, "250,000 allied soldiers stormed France's Mediterranean shores on August 15, 1944. 70 days after the D-Day landings at Normandy, catching German troops in a pincer so tight that Hilter muttered to aides, "This is the darkest day of my life." Operation Dragoon was to have coincided with the June 6, Normandy assault, but there were no landing craft to spare. When it finally happen Paratroopers bore much of the brunt. Breuer (1997) reported, "It was a complete blackout as the gigantic sky train carrying initial elements of General Bob Frederick's First Airborne Task Force barrowed through the calm Mediterranean night. Hanging in majestic aloofness in the black heavens was a sliver of a crescent moon." The official after action reports described by Retired Colonel Tom Cross of the 517th PRCT are available to read at the 517th web site that is www.517prct.org. This is his report of Operation Dragoon:

The Airborne Invasion of Southern France
Operation Dragoon

Preface

Our generation, which was the first to employ parachute and glider borne troops in armed combat against an enemy force, should refuse to accept revisionist changes in the description or accounts of the deployment of these forces and their subsequent operations should they differ from what we actually experienced.

If we are not vigilant in the preservation of the history of our actions, then subsequent generations of historians might try to color the facts with their own differing interpretation of who we were and what we did. This account of OPERATION DRAGOON will, for the most part, deal with official After Action Reports written by airborne units, and their higher echelons of command, that were directly involved in the planning and conduct of OPERATION DRAGOON.

This historical account will primarily present the big picture aspects of OPERATION DRAGOON. Most Airborne and Troop Carrier units that participated in this operation have already published their own individual unit histories which include their specific roles and actions. What is presented here will compliment these unit histories. When specific Unit Reports are used they are intended to fill out and amplify the broader aspects of OPERATION DRAGOON.

Background

The initial concept and planning of OPERATION DRAGOON was shrouded in controversy beginning as far back as 1943. It was during the QUADRANT conference, held in Quebec during August 14-24, 1943, that both President Roosevelt and Prime Minister Churchill, their advisors, and the Combined Chiefs of Staff, decided on a target date for OPERATION OVERLORD, the invasion of Normandy. The date for OVERLORD was initially set for May 1, 1944. At this same time, the invasion of Southern France was decided on and the code name of ANVIL was assigned to the operation.

Prime Minister Churchill, even at this very early date, voiced his concern over the selection of the Southern France approach and strongly championed a military inroad into the Balkans. Churchill resisted ANVIL (the code name later changed to DRAGOON) right up until August 9, 1944, only six days before the actual invasion of Southern France. The change in code name from ANVIL to DRAGOON mischievously reflected Prime Minister Churchill's view that he was pressed by the Americans into doing something that did not enjoy his full support.

The EUREKA Conference in Teheran (from November 28th to December 1, 1943) between Marshall Stalin, Prime Minister Churchill and President Roosevelt once again confirmed that there would be a landing in Southern France. The exact details of the plan were left in limbo due to the still present opposition by Churchill. However, Roosevelt did insist that both OVERLORD and ANVIL (now DRAGOON) would receive top priority in future plans.

The introduction of additional combat and support units to the troop list for OVERLORD as well as the current state of operations in Italy indicated that DRAGOON would have to be executed at a date later than that set for OPERATION OVERLORD. The difficult ANZIO OPERATION caused 68 LST's to be diverted from resources previously OVERLORD. Thus the shortage of LSTs for the OVERLORD seaborne lift and necessity for maximum Troop Carrier resource for the airborne phase of OVERLORD dictated a change to the invasion sequence as originally developed at earlier strategic planning conferences.

General Eisenhower was a strong supporter of DRAGOON, even if the operation had to be pushed off to a later date. His primary concern was the achievement of a military victory in Europe in the shortest possible period of time. DRAGOON offered General Eisenhower the opening of the valuable Port of Marseilles, the second largest port in France, plus the chance to open up a second front in Europe, the practical outcome of which was to detain a large number of German forces in the Southern France theatre of operations and preventing them from reinforcing their forces in Normandy. Another important factor in Eisenhower's arguing for DRAGOON was that it gave the Free French Forces a meaningful and important role in the liberation of their homeland.

It becomes apparent, then, the political versus military controversy surrounding what is perhaps one of the best planned and well executed invasion operations conducted during WWII. The success of the airborne phase of DRAGOON came about through the military professionalism and leadership displayed at all levels of staff and command from the highest headquarters and command level right down to the squad and initiative of the individual airborne trooper.

Despite a last minute plea from Prime Minister Churchill, on August 8, 1944, the United States Joint Chiefs of Staff cabled General Eisenhower that OPERATION DRAGOON would proceed as planned. This was the "green light and go" for the 1st Airborne Task Force (ABTF), a decision that played itself out over the Drop Zones and Landing Zones of Southern France.

Introduction

Initial planning for the Airborne operation in DRAGOON was begun by the planning staff of the Seventh Army in February 1944. The status of airborne units in the Mediterranean Theater at that time materially influenced the planning at this stage as none of the Airborne or Troop Carrier Wings were fully prepared to undertake airborne operations.

The 51st Troop Carrier Wing, composed of three groups, had remained in the Theater after the inactivation of the XII Troop Carrier Command. However, only a portion of the 51st wing was available for airborne training because of the demand for Troop Carrier aircraft for special operations, air evacuation and general transport requirements. A few of these aircraft were intermittently attached to the Airborne Training Center located in Trapini, Sicily and later at the new location of the Airborne Training Center at Lido de Roma for a program of limited airborne training. At the Center, the First French Parachute Regiment, two Pathfinder Platoons and the American replacements received limited airborne training.

The British 2nd independent Parachute Brigade Group commanded by Brigadier C.H.V. Pritchard, Batteries A & B of the 463rd Parachute Artillery battalion commanded by Lt. Col. John Cooper and the 509th Parachute Infantry battalion commanded by Lt. Col. William P. Yarbrough were withdrawn from the front in Italy and given intensive training with a full Troop Carrier Group of the 51st Wing. The War Department was requested to provide an airborne division for employment in DRAGOON, but in lieu of this, a number of separate airborne units were shipped to the Theater. These were the 551st Parachute Infantry Battalion commanded by Lt. Col. Wood G. Joerg, the 517th Parachute Regimental Combat Team (PRCT) commanded by Colonel Rupert D. Graves and the 550th Glider Infantry Battalion commanded by Lt. Col. Edward I. Sachs. The 517th consisted of three parachute infantry battalions: the 1st Battalion commanded by Lt. Col. William J. Boyle, the 2nd Battalion commanded by Lt. Col. Richard J Seitz and the 3rd Battalion commanded by Lt. Col. Melvin Zais. The 460th Parachute Artillery Battalion of the 517th PRCT was commanded by Lt. Col. Raymond L. Cato and the remaining contingent of the 517th PRCT which was the 596th Parachute Combat Engineer Company which was commanded by Captain Robert W. Dalrymple. The 517th PRCT arrived from the States after just having completed its final large scale airborne training exercise and participation in the Tennessee Maneuvers. The 517th PRCT was attached to the fifth Army for ten days of battle experience in the lines and the 551st and 550th Battalions were attached to the Airborne Training Center, and then located in Sicily for training.

Thus, by the middle of June 1944, there were considerable airborne forces in the Theater which could be considered for airborne operations. In order to secure the utmost cohesion and to obtain the optimum results, it was decided to move the Airborne Training Center with its attached units, as well as the troop carrier aircraft (now increased to two full groups) of the 51st Troop Carrier Wing to the Rome area. Here was established a compact forward base for all airborne forces in the Theater.

Organization
Airborne Elements

Toward the first of July, 1944, the plans for OPERATION DRAGOON were made firm, including the use of a provisional airborne division made up of available units in the Theater. Major General Robert T. Fredrick, formerly commander of the First Special Service Force and later the commander of the 36th Infantry Division, assumed command of the composite force. Conferences were held immediately to secure the additional supporting units needed to organize such a balanced airborne force. Certain units on the DRAGOON troop list were earmarked for this purpose. Authority was requested from the War Department to activate those units which were not authorized on the Theater Troop List. By July 7th initial instructions relative to the organization of the Provisional Airborne Division known officially as the Seventh Army Airborne Division (Provisional) was disbanded and the 1st ABTF was constituted by a secret Adjutant General order of July 18, 1944. Assignment was to the North Africa Theater of operations and activation was accomplished by the reassignment of personnel from the Seventh Army Airborne Division (Provisional). This action was concurrent.

The following units which just a short time before had been assigned to the Provisional Airborne division, were then assigned to the 1st ABTF:

1. 2nd British Independent Parachute Brigade Group
2. 517th Parachute Infantry Regiment
3. 509th Parachute Infantry Battalion
4. 550th Airborne Infantry Battalion
5. 1st Battalion, 551st Parachute Infantry Regiment (Reinforced)
6. 460th Parachute field Artillery Battalion
7. 463rd Parachute Artillery Battalion
8. 602nd Pack field Artillery Battalion
9. 596th Parachute Combat Engineer Company
10. 887th Airborne Aviation Engineer Company
11. 512th Airborne Signal Company
12. Anti Tank Company 442nd Infantry Regiment
13. Company A, 2nd Chemical Battalion (Motorized)
14. Company D, 83rd Chemical Battalion (Motorized)
15. 172nd DID British Heavy Aerial Resupply Company
16. 334th Quartermaster Depot Company, Aerial Resupply
17. Detachment, 3rd Ordnance Company (MM)
18. 676th Medical Collecting Company (Designated 07/29/44)
19. 5 Unit Pathfinder Platoons (These were unauthorized but were formed from 1st ABTF resources)

The 1st ABTF was then given a 5 per cent over strength by the assignment of parachute filler replacements from the Airborne Training Center. Meanwhile, the activation of the Task Force headquarters and headquarters Company, two additional batteries of the 463rd Parachute Artillery Battalion, the 512th Airborne Signal Company, and an anti tank unit to be designated the 552nd Anti Tank Company for operations because of the short time remaining before D-Day. The 442nd Anti Tank company was well trained prior to arrival in the Theater and when it was necessary to re-equip them with the British six-pounder, since the 442nd's 57mm anti tank gun would not fit into the CG-4A glider, this well trained unit was able to make the transition in record time.

Due to the shortage of qualified airborne officers in the Theater, it was necessary to ask the War Department to make available a divisional staff for General Fredrick. Thirty-six qualified staff officers arrived in the Theater by air toward the middle of July. Most of these came from the 13th Airborne division and a few from the Airborne Center at Camp Mackall, North Carolina.

Certain other organizations were made available to the 1st ABTF on the basis of their being employed in the preparatory stages but not in the operation itself. These units included detachments from a Signal Company, a Quartermaster Truck Company, and some 400 replacements from the Airborne Training Center.

Concentration of Units

As previously described, the Airborne Training Center and the 51st Troop Carrier Wing had been ordered to the Rome Area and established a compact airborne base at Ciampino and Lido de Roma Airfields. By the 3rd of July, and advance echelon of the Airborne Training Center was established at the Ciampino Airfield and was ready to operate. By the 10th of July, the Center with its attached units, the 551st Parachute Infantry Battalion, and the 550th Glider Infantry Battalion were completely located at the Airborne Base. The Divisional Staff ordered from the United States for the 1st Airborne Task Force Headquarters could not arrive until approximately the 15th of July. Therefore, all other American units in the Theater were attached to the Airborne Training Center so that its staff could be used to assist I the expediting of the concentration of airborne troops. The 517th PRCT was ordered out of the line from Fifth Army control and arrived in the Rome Area by July 5th, 1944. The 509th PIB, already located at Lido de Roma was similarly attached to the Airborne Training Center for instruction. The various supporting arms and services which had been placed at the disposal of the Seventh Army Airborne Division (Provisional) were similarly attached. By July 17, 1944, General Fredrick had moved his Headquarters to Lido de Roma and was ready to proceed with the final organization and training of the airborne forces and the detailed planning of OPERATION DRAGOON. On the 21st of July, General Fredrick requested that the name of the Provisional organization be changed to the '1st ABTF, since the use of the term "Division" was considered a misnomer.

Planning

Although continuous planning for DRAGOON had been under consideration since February 1944, no final detailed planning was possible until the 1st Airborne Task Force and the Provisional Troop Carrier Air Division were organized and prepared to function. As of consequence of this, the final planning could not commence until almost the 20th of July. Upon his arrival, General Williams, the Commanding General, Provisional Troop Carrier Air division, approved the suggested plan of utilizing the previously selected Rome area as a training site. He also concurred in the choice of the previously selected take-off airfields located at Ciampino, Galera, Marcigliano, Fabrisi, Viterbo, Tarquinia, Voltone, Montalto, Canino, Orbetello, Ombrone, Grossetto, Fallonica, and Piombino. Subsequently, the Provisional Troop Carrier Air Division undertook primarily the planning aspects of the operation involving high level coordination, timing, routes, corridors, rendezvous, and traffic patterns. In general, planning for the details of the selection of Drop Zones (DZs), Landing Zones, (LZs) and composition of lifts were left to the Airborne and Troop Carrier units involved.

It was first decided that a pre-dusk airborne assault on D-Day should not be made, as this might jeopardize the success of the entire operation. Second, it was decided that it would not be necessary or advisable to launch the initial vertical attack after the amphibious assault had begun. The latter decision was reached in view of the wide experience of our Troop Carrier crews in night take-off operations, and because of the marked improvement that had been made in the pathfinder techniques. Consequently, the basic plan called for a pre-dawn assault. One proposed plan contemplating an immediate staging in Corsica was rejected because of lack of available Corsican airfields, and also because those fields available were located on the eastern side of the island, and their use would have necessitated a flight over 9000 foot mountain peaks. Such a flight would be difficult even for unencumbered transport aircraft, and for C-47's towing gliders it was considered excessively dangerous. A further consideration was the fact that such an intermediate staging would have required the establishment of the airborne corridor south of the main naval channel, and would have necessitated the adoption of a dog-leg course for the flight.

After several conferences had been held at Seventh Army Headquarters, with all Army, Navy, Air and Airborne commanders concerned, the rough plan was drawn up and approved about July 25th, 1944. This plan envisaged the use of an airborne division prior to H-Hour with the dropping of the airborne pathfinder crews beginning at 0323 hours on D-Day. The main parachute lift of 396 plane loads was to follow, starting at 0412 hours and ending at 0509 hours. The first glider landing was to take place at 0814 hours and continue on through until 0822 hours. Later in the same day a total of 42 paratroop plane loads were to be dropped, followed by 335 CG-4A gliders starting at 1810 hours and ending at 1859 hours. The automatic air resupply which was to have part of the D-Day late afternoon mission was postponed at a late stage of the planning processes because of insufficient Troop Carrier aircraft and because Troop Carrier Command would not drop supplies from aircraft towing glider in the afternoon glider lift. The final plan provided for 112 plane loads to be brought in automatically on D+1. The remaining supplies were to be packed and held available for emergency use by either the 1st ABTF or by any Seventh Army unit which might become isolated.

The Troop Carrier routes selected were carefully chosen after due consideration of the following: the shortest distance, prominent terrain features, traffic control for the ten Troop Carrier Groups, naval convoy routes, position of assault beaches, primary aerial targets, enemy radar avoidance, excessive dog legs, prominent landfalls, and position charts of enemy flak installations. This route logically followed the Italian coast generally from the Rome area to the island of Elba, which was used as the first over water check point, followed by the tip of the island of Corsica and proceeded on an azimuth course over Naval check points to the landfall just north of Frejus and Agay. Complete plans were made with the navy on the position of this airborne corridor, and detailed information concerning it was widely disseminated among the naval forces.

Because of high terrain features in the target area it was decided that it would be necessary to drop the paratroopers and release gliders at exceptionally high altitudes varying from 1500 to 2000 feet. Glider speeds for towing were set at 120 M.P.H. and dropping speed 110 M.P.H. the formation adopted for the parachute columns was the universal "V of Vs" in nine ships in serials of an average of 45 aircraft, with five minute intervals head to head between serials. The glider columns adopted a "pair of pairs" formation echeloned to the right rear. Serials made up of 48 aircraft towing gliders in trail were used with eight minute intervals between serial lead aircraft. Parachute aircraft employed a maximum payload of 5430 lbs., Horsa gliders 6900 lbs., and the CG-4A gliders a total of 3750 lbs.

Difficulty in the procurement of maps and models proved to be a serious inconvenience in the planning and preparations for the operation. Map shipments in many instances were late in arriving or were improperly made up. Terrain models on a scale of 1:100,000 were available but the most useful terrain model, a photo-model in scale of the 1:25,000 was available only in one copy, which was wholly inadequate to serve both the Provisional Troop Carrier Division and the 1st Airborne Task Force. The blown-up large scale photographs of the DZ-LZ areas in particular were excellent, but these arrived too late for general use. The original coastal obliques were not of much assistance to the Provisional Troop Carrier Air Division since the run-in form the IP (First landfall) was not adequately covered. These late photographic studies uncovered the previously unknown element of anti-glider poles installed on the LZs. All earlier photo studies had failed to reveal this pertinent information. An excellent terrain model was turned out by the 2nd British Independent Parachute Brigade and it was of great assistance to that unit for the operation.

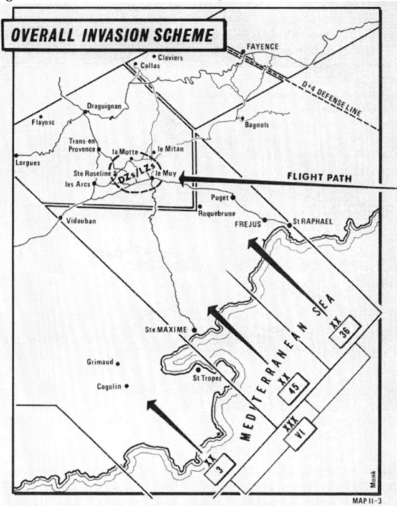

Figure 1: Overview of Overall Invasion Plan
Map drawn by Frank (Monk) Huffman

Pre-Operation Training

By the middle of July nearly all the airborne units to be employed on OPERATION DRAGOON had been assembled in the Rome area. An intensive final training program had begun by the 1st ABTF in conjunction with the Airborne Training Center. Of the Airborne units to be used in the OPERATION, only the 509th Parachute Infantry Battalion and the 2nd British Independent Parachute Brigade Group had received any recent combined airborne training with the Troop Carrier units. The 517th PRCT had just come out of the line with the Fifth Army as had the 463rd parachute Artillery Battalion. Other units, such as the 551st Parachute Infantry Battalion and the 550th glider Infantry Battalion had but recently arrived overseas and had been given a course in ground and airborne refresher training at the Airborne Training Center.

A particularly urgent task was the training of the newly chosen glider borne troops. A combined glider school was set up and instruction commenced in the mechanics of loading and lashing for the units involved. The units involved in this difficult last minute procedure were the 602nd Pack Field Artillery Battalion, the 442nd Infantry Anti Tank company, the 887th Airborne Aviation Engineer Company, Company A, 2nd Chemical Battalion, Company D, 83rd Chemical Battalion, and the various other units such as the Division Ordnance Detachment and the Medical Collecting Company. Once these troops had finished the course in loading and lashing, they were given orientation flights and finally one reduced size practice operational landing on a simulated LZ.

The Provisional Troop Carrier Air Division Pathfinder unit went to work with the three airborne pathfinder platoons and thoroughly tested the radar and radio aids to be used in the operation. This training was divided into three phases, the first being concerned with the technical training with "Eureka" sets, M/F beacons, lights and panels. Tests were made to locate any deficiencies in either the training or the equipment to be used. The second phase was devoted to practice by the crews in using the equipment as a team. All the teams practiced in setting up and operating the equipment under all possible conditions. The third phase emphasized actual drops with full equipment in which every attempt was made to secure the utmost realism in the preparatory exercises. Small groups of parachute troops were dropped on the prepared DZs to test the accuracy of the pathfinder aids.

Due to the lack of time and the difficulty of re-packing the parachutes in time for the operation, it was impossible for the 1st Airborne Task force to stage any large scale realistic final exercise. The various individual units did participate in practice drops to the fullest possible extent, generally using a skeleton drop of two or three men to represent a full "stick" of paratroopers, the reminder of the unit already being on the DZ so that the assembly procedures could be tested and experienced. A combined training exercise with the Navy, however, was scheduled. All naval craft carrying water-borne navigational aids were placed in the same relative positions as in the actual operation. A token force of three aircraft per serial were flown by all Serial Leaders over these aids, the flights flown on the exact timing schedules, routes and altitudes as were used in the operation. Two serials of 36 aircraft each were flown over this same route in the daylight in order that the Naval Forces would become acquainted with the troop Carrier formations. Further practice runs were made by the Troop Carrier units in conjunction with the 31st and 325 fighter Groups so as to work out the details of the fighter Cover Plan and the Air/Sea Rescue Plan.

In view of the fact that the Task Force was composed of units that had not previously worked together, training of combat teams, as organized for the operation, was emphasized to further successful operations after landing. Training of each newly organized combat team was conducted on terrain carefully selected to duplicate, as nearly as possible, the combat teams sector within the target area.

The problem of securing and organizing qualified personnel and then training those units that had to be activated on short notice proved to be difficult. Such highly specialized personnel as are required for an Airborne Signal Company or for an Airborne Divisional Headquarters were extremely difficult to find in an overseas theater. Consequently, these personnel had to be located in the local Replacement Depots or at the Airborne Training Center and then trained for the specific positions they were to fill. The highest praise is due General Fredrick and his staff, and the Airborne Training Center for the manner in which this task was accomplished.

It was fortunate that the larger elements of the Task Force, particularly the Combat Teams, were already well trained. Some of them were battle-seasoned and nearly all were accustomed to providing for themselves since each was basically designed to be a separate regiment, battalion or company. This unique circumstance allowed the units not only to look after their own requirements, but also permitted them to aid the 1st ABTF as a whole during this period of training.

The Operation - Airborne Phase

The night of D-1 was clear and cool in the take-off areas used by the Airborne Forces in DRAGOON. The Troop Carrier units were at their stations in the ten airfields extending from Ciampino near Rome to Fallonica, north to Grosseto. Due to the serious lack of ground transportation, it was necessary for the bulk of the Task Force, except for the British 2nd Independent Parachute Brigade Group, to commence the movement to the dispersal airfields by D-5. By D-2, the Airborne Forces had been shuttled from their training and concentration areas near Rome to their designated airfields. The C-47 aircraft had all been deadlined, checked thoroughly and were in excellent condition for the invasion flight. All of the preliminary checks had been completed at the glider marshalling airfields and they were ready to roll. A feeling of assurance as to the outcome of the operation prevailed among all elements.

As to be expected in any airborne operation, prevailing weather was to be important and could influence the parachute drops. Once the target date had been set it could not be changed for the benefit of the Airborne Forces, even though it necessitated a drop with out the assistance of moonlight. It had been hoped that the drop could be made on a clear night so that the Troop Carrier aircraft could identify large hill masses and coastal features as possible check-points. However, on August 14-15, 1944, all of Western Europe was covered by a large flat high pressure area centered over the North Sea. A portion of this "High" had broken off and had settled over the main target area. This did preclude the probability of any sizeable storm or heavy winds, and the only threat was one of accumulating fog or stratus. Consequently, the forecast for the operation was clear weather to Elba, followed by decreasing visibility until the DZs were reached, at which time the visibility was expected to be from 2 to 3 miles. Actually, the haze was heavier than anticipated and the visibility was less than half a mile on the DZs. The valley fog that had completely blanketed the early parachute operation had dispersed by 0800 hours. Fortunately, this was in time for the morning glider mission. Considerable navigational difficulties arose from the fact that the wind forecast was almost 90 degrees off the direction initially indicated. Consequently, the navigators could make necessary corrections only by use of check-points over the water route. Fortunately, the wind did not reach high velocity and was less than six M.P.H. on the DZs.

The operation was prefaced by a successful airborne diversion designed to serve two purposes in the Cover Plan. First, it was to create the illusion of a southern airborne corridor; second, it was to simulate a false airborne DZ by dropping rubber parachute dummies into selected areas. The six aircraft used on this mission dropped "window" enroute to give the effect of a mass flight and at 0205 hours on D-Day they dropped 600 parachute dummies as panned on false DZs located north and west of Toulon. German radio reports indicated the complete success of this rouse. The rifle simulators and other battle noise effects used in the diversion functioned well and added to the realism of this feint.

The Airborne operation began shortly after midnight on August 14-15, 1944. Aircraft were loaded, engines were warmed up and the marshalling of aircraft for takeoff was underway at 0300 hours. At the same hour as the first Troop Carrier aircraft took off with their load of three Pathfinder units. During the aircraft marshalling phase several aircraft received minor damage and one was demolished. One aircraft from the 439th troop Carrier Group crashed and burned on take-off, two aircraft struck trucks and received minor damage and two other aircraft from the same Group had a collision while rolling out on the taxi ramp. One other aircraft suffered damage to a wing when one of its wheels hit a poorly filled hole on the airfield, and another suffered damage due to the premature release of a parapack. Considering that the take-off was from prepared landing strips without any moon to aid the poor visibility caused by excessive dust, the success of this phase of the operation is unquestioned. An estimated eight glider had difficulty on take-off and had to be unloaded so that the substitute gliders could be employed to lift the loads.

The Pathfinder Mission is of interest in that instead of the reported "complete success" of the mission, later facts revealed that it was less than fifty per cent effective. Nine Pathfinder aircraft divided into three serials were each employed in such a manner that three teams would be dropped on each DZ. The Lead Serial, supposed to drop Pathfinders from the 509th and 551st Parachute Infantry Battalions plus that of the 550th glider Infantry Battalion at 0323 hours lost its way and circled back to sea to make a second run a the correct DZs. After circling for about half an hour one aircraft dropped its Pathfinder Team and went home. The other two aircraft separated after that and one dropped its team about 0400 hours. The last Team jumped at 0415 hours on the sixth run in on the target. Lost in the woods and far from their objective, none of the three Teams were able to reach the Le Muy area in time to act as Pathfinders. The Second Serial was to drop its Pathfinders from the 517th PRCT at 0330 hours. The actual drop was at 0328 hours. The 517th Pathfinders landed in the woods three and a half miles east of DZ "A" and just east of Le Muy itself. A two minute jump delay would have put them right on target. To make matters worse they were attacked and spent time beating off the assault.

They arrived at DZ "A" at approximately 1630 hours and set up a Eureka Beacon, a MF Beacon, and a "T" Panel which was of assistance to the afternoon missions for this area. The Third Serial carrying the Pathfinders form the British 2nd Independent Brigade Group, dropped on DZ "O" at 0334 hours, exactly on schedule. By 0430 hours they had two Eureka Beacons in operation about 300 yards apart along the axis of the approach. They also set up two lights for assembly purposes. This Pathfinder drop was the most accurate of the three.

Approximately one hour after the Pathfinder aircraft took off the main parachute lift, composed of 396 aircraft in nine serials, each averaging 45 aircraft, took off and proceeded on their assigned courses. The flight toward the designated DZs was marked by no untoward incident. Amber downward recognition lights were employed until the final water checkpoint had been crossed. Wing formation lights were similarly employed and no instance of friendly naval fire on the Troop Carrier aircraft was reported. No enemy aircraft was encountered during the flight. Of particular note is the fact that over four hundred Troop Carrier aircraft had flown in relatively tight formation under operational strain for some five hundred miles without incident. The many hours of time devoted to training in night formation flying had produced excellent results.

The radio, radar and other marker installations undoubtedly helped to save the day in terms of the success of the mission. The Eurekas which had been installed at each Troop Carrier Wing Departure Point, the Command Departure Point, the North East tip of Elba, Giroglia Island (North Corsica), on the three marker Beacon boats spaced 30 miles apart on curse from Corsica to Agay, France (the first landfall checkpoint), all worked exceedingly well with an average reception range of 25 miles. Holophane lights similarly had been placed at these positions and did aid the navigators in their work with the contrary wind currents. Their reception was an average of 8 miles until the DZs were reached, at which time they became invisible because of the haze and ground fog. MF Beacons (Radio Compass Homing Device) were installed at Elba, North Corsica and on the central oat marker beacon and dropped on the DZs along with the Eurekas and the Holophane lights. Many pilots reported that they picked up these signals up to 50 miles away and often kept the aircraft on beam when they occasionally lost the Rebecca signal on their Eureka, which in to many cases exhibited a tendency to drift off the frequency despite constant operational checking. It should e emphasized again that the entire parachute drop was made "blind" by the Troop Carrier aircraft who had to depend on these MF Beacons and Eureka sets for their signal to drop the paratroopers. Brigadier C.H.V. Pritchard, Commanding Officer, 2nd British

Independent Parachute Brigade Group felt that this single deficiency could have jeopardized the complete operation.

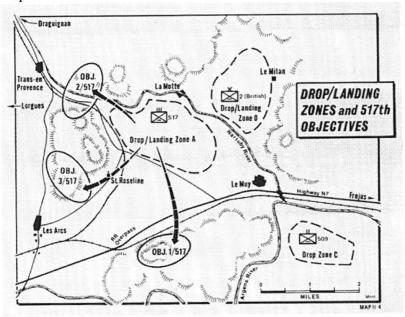

Figure 2: Initial 517th PRCT D-Day Drop/Landing Zones and Objectives
Map drawn by Frank (Monk) Huffman

Initial reports stated that the parachute drop was eminently successful and it was given a score card rating of 85 per cent accuracy. This was later down graded as the incoming reports became more accurate and detailed in content. Despite the accuracy of the reporting system sufficient paratroopers landed on the DZs or in the immediate vicinity thereof, in areas which for all practical purposes can be considered as contiguous to the DZs and from which terrain the parachute forces were in positions which allowed them to carry out their assigned missions. This was accomplished despite the handicaps of no moon, general haze, and heavy ground fog. An estimated 45 aircraft completely missed their designated DZs. Some of these dropped their paratroopers as far as 20 miles from the selected areas.

Among the aircraft that missed the DZs were twenty in serial Number 8, which released their troops prematurely on the red light signal. The only plausible explanation that can be offered is that a faulty light mechanism in one of the leading aircraft must have gone on green prematurely and the troops in the lead aircraft jumped according to this signal. The paratroopers in the following aircraft, on seeing the leading aircraft's troopers jump, probably did likewise and jumped even though the red signal still showed in their won aircraft. This group was principally comprised of elements of the 509th Parachute Infantry Battalion and about half of the 463rd Parachute field Artillery Battalion. Two "sticks" of paratroopers landed in the sea off of St. Tropez, near Cannes. The remainder made ground landings in the vicinity of these two towns. Although far from the designated DZ, these elements organized themselves, made contact with the French Resistance Forces, and proceeded to seize and hold St. Tropez. Approximately 25 aircraft from another Troop Carrier Group mistakenly dropped their paratroopers some 15 miles north of Le Muy near Fayance. The troops in this instance comprised part of the 5th Scots Parachute Battalion of the British 2nd Independent Parachute Regimental Brigade Group, and elements of the 3rd Battalion, 517th PRCT. Although some 20 miles from their DZ, these troops either undertook individual missions or fought their way back to their own units in the proper Objective Area. By evening of D-Day, most of this group was reassembled on DZs "A" and "O". The Task Force

Chief of Staff, along with the Task Force Surgeon and other key staff officers were among this group. DZ "A" generally west of Le Muy had a tendency, during the drop, to become merged with DZ "O", slightly northwest of this key town on the Vargennes Valley, which caused considerable confusion later on in the day. This inadvertent merging of the two DZs also produced some confusion and difficulty during the period of the equipment bundle recovery. This confusion was further increased because the British 2nd Independent Parachute Brigade Group was using different equipment from that of the 517th PRCT on DZ "A".

The terrain of the DZs on which the paratroopers landed was in general excellent for such an operation. DZs "A" and "O" covered an area of small cultivated farms consisting mainly of vineyards and orchards. There were very few large buildings, telephone wires, tall trees and other formidable obstacles. The anti-airborne poles established in the Parachute Drop Areas were not sharp or placed in sufficient density to obstruct the parachute landings to any material degree. A total of 175 paratroopers, scarcely more than 2 per cent, suffered jump casualties. DZ "C" on which the 509th Parachute Infantry Battalion Combat Team jumped was a hill mass more rugged than the ground of the other DZs, but even this rougher terrain did not interfere with the success of the jump.

Serial Number 14 (the first of the glider serials) made up of supporting artillery and anti tank weapons for the British 2nd Independent Parachute Brigade Group departed as scheduled for its 0800 hours glider landing, but was recalled because of heavy overcast. The flight circled for one hour and landed at 0900 hours. One glider and tug aircraft had to turn back. One glider ditched off shore and another disintegrated in mid-air over the water (the cause was subsequently laid to structural failure0. The stakes driven into the ground all over the LZs did not prove to be difficult obstacles, even though the poles did cause considerable damage to the gliders and in some instances to their loads. The anti-glider poles served in many instances as additional braking power for the gliders, since the poles were small, planted at shallow depth, and were too widely dispersed to perform their intended mission. Evidently, the French farmers who were forced to plant the anti-airborne (glider) stakes had done the minimum work they could in this forced construction. These poles were on the average of 12 feet high and 6 inches in diameter. They had been driven in the ground less than two feet and were generally more than 30 to 40 feet apart. Serial Number 16 was a parachute load made up of the 551st Parachute Infantry Battalion. It dropped accurately on DZ "A" at 1800 hours as planned. This drop was followed up rapidly by continuous glider serials Numbered 17 through 23, nine gliders was reported to have been released prematurely, four of which made water landings. A large percentage of their crews and personnel were saved by prompt action on the part of the Navy. The landing skill of our glider pilots was outstanding. Although the 1000 foot interval adopted for towing caused considerable jamming over the LZs, these pilots affected excellent landings. Several pilots ground-looped their gliders to avoid obstacles and still brought in personnel and cargoes safely. Another reason for the crowded conditions over the Landing Zones was a notable tendency on the part of successive flights to seek additional altitude as a result of the 'accordion movement" of the flights enroute. In turn, this progressively created a greater mass of aircraft and gliders being over the LZs at any one time than had been contemplated. Further difficulty arose because the pilots of the early glider lifts landed on the best and most obvious sections of the Landing Zones instead of in their own designated sectors. On their arrival, the later lifts consequently found that their assigned landing areas were almost entirely occupied with gliders which forced them to seek alternate and less desirable areas. This was further compounded by the situation wherein two glider serials were released so close together that they both were in the air at the same time over the same already crowded LZ. Quick reaction was the order of the day for one did not get a second chance to make a landing run in a glider. The pilots simply had to dig in on their landings because of the limited space. These abrupt, heavy landings, did cause excessive damage to the gliders, primarily because of the lack of the 'Griswold Nose" modification. The glider pilots demonstrated great presence of mind, prompt action and skillful maneuver saving many lives and much valuable airborne equipment. It was established by D+6 that not more than 125 glider-borne personnel were injured in these landings. Although not encountered in the immediate objective area, as a matter of interest, it is worthy to note that in the Frejus area outside of the Drop Zones, there was a second type of anti-glider obstacle which consisted of small but sturdy sharpened stakes some 18 inches high, firmly imbedded in the ground and connected by wire which could play havoc on the belly of any glider landing on terrain prepared with such an obstacle.

In general, the problem of air re-supply did not become as urgent as had been expected. Absence of serious enemy opposition caused ammunition expenditure to fall below the anticipated amount. The initial plan for bringing in the first supplies by air on D-Day was consequently changed so that it was not until 1100 hours on D+1 that two Troop Carrier Groups brought 116 aircraft loaded with supplies. The aircraft arrived over the DZs on schedule but at an altitude well over 2000 feet, which made accurate dropping extremely difficult. A rather stiff breeze, coupled with the high altitude and the merging of the DZs ("A" and "O"), caused much of the dropped equipment and supplies to get into the hands of the wrong units on the ground. Well over 95 percent of the 1700-odd bundles dropped by parachute landed safely, but much of the specialized equipment failed to reach the units which had requested it. Subsequently, re-supply missions carrying emergency signal and medical supplies were flown again on the night of D+1. Although these drops had to be made at night by pathfinder aids the success of these missions was above average, except that again the high altitude caused excessive scattering of equipment. The 334th quartermaster Depot Company, Aerial Re-supply, aided by the Parachute Maintenance Section of the 517th PRCT, packed over 14,000 parachutes and 1000 tons of equipment for the operation and deserved commendation for the outstanding work they accomplished. The British Allied Air Supply Base deserves the very same credit and a well done for their contribution.

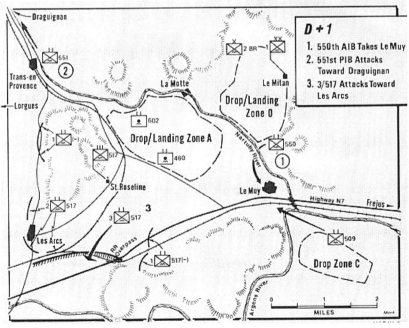

Figure 3: D-Day +1
Map drawn by Frank (Monk) Huffman

Ground Phase

While in the grand scheme of the whole operation, enemy strength and actions could be considered light, it was a far different matter from the troopers standpoint as he and his comrades, in individual actions and small unit combat situations, struggled to get back to their designated DZs and Assembly Areas, from which they had been separated, due to a 50 percent dropping error during the final run into the Objective Area. To these "mission oriented" troopers this was in every sense of the word a "big war". How the Airborne Forces accomplished their missions under adverse conditions can best be illustrated by taking the largest airborne unit involved during this phase of the operation, the 517th PRCT, and highlight its D to D+1 combat actions.

The 3rd Battalion of the 517th PRCT dropped approximately 15 to 20 miles northeast of DZ "A" and was scattered over 8 miles of rough terrain from Élans east through Fayence to Tourettes and Callian, all a good days march from DZ "A". As this unit began to assemble and look for its equipment, three major sub-sections of this unit emerged; the first sub-section consisted of the first 10 aircraft loads dropped near Seillans which included most of Company I, the Battalion Commander, plus a Battalion headquarters contingent all totaling 160 troopers; the second contingent consisted of 60 troopers from the Battalion Headquarters Company and G and H Companies who dropped in the vicinity of Tourettes; and finally, the third group of over 200 troopers from units within the Battalion plus members from the 596th Parachute Combat Engineer Company and members of Regimental Headquarters and Service Companies who had all been attached to the 3rd Battalion for the drop. This entire last group had been dropped in the vicinity of Callian. The combined strength of these sub-sections totaled about 480 troopers. About 35 troopers were injured during the drop and had to be left behind, with an escort, due to the nature of their injuries. Seventy-five troopers caught up later and about 50 others were too far away to join up and so they resorted to independent actions of opportunity. By 0800 hours the Battalion Commander, Lt. Col. Melvin Zais, with almost all of his Battalion now intact commenced to march to be back to his Objective Area which was some distance away.

The 1st Battalion, 517th PRCT was also dropped in a scattered pattern over an area 30 to 40 square miles west to northwest, and southwest of trans-en-Provence. At daylight Captain Charles La Chaussee of Company C and Lt. Erle Ehly of Battalion Headquarters Company, had managed to get together about 150 troopers in the Battalion Assembly Area and had made the decision to move out to the Battalion Objective Area without further delay. At this point, the Battalion Executive Officer, Major Herbert Bowlby, joined them and took command. The Battalion Commander, Major William Boyle, landed about 4 to 5 miles from Trans and having lost about two-thirds of his "stick" in the dark gathered the remaining 5 troopers and took off for his Objective Area. He entered the outskirts of Les Arcs in the afternoon and found and additional 20 troopers that included the Battalion S-3 and the Assistant Battalion Surgeon. Gathering up this group, Major Boyle again started out for the Objective Area. On the way he and his party got into a firefight with about 300 of the enemy and the situation became "touch and go". Additional 1st Battalion troopers coming in from the northwest joined Major Boyle until his force numbered 50 or more troopers. He was forced to set up a perimeter type defense on the southeast edge of Les Arcs and dig in. A third group from the same Battalion managed to meet up with Captain Donald Fraser who's Company A had been designated as the 517th PRCT Reserve Force. Captain Fraser managed to get together about 200 troopers to take over the Area of operations that was previously assigned to the 3rd Battalion, 517th PRCT on the west of the Chateau Saint Rosaline; and Major Boyle encircled in the vicinity of Les Arcs.

The 2nd Battalion, 517th PRCT, commanded by Lt. Col. Richard Seitz, at about 50 percent strength, was on its way to its Objective Area by daylight after having landed about a mile from its Drop Zone, which was DZ "A". By noon of D-Day the 2nd Battalion minus Company F was on its Objective. Company F which had dropped with part of the Regimental headquarters and part of the 596th Parachute Combat Engineer Company in the vicinity of Le Muy rejoined the 2nd Battalion, 517th PRCT on the Battalion Objective later in the afternoon.

The versatility of the airborne troops was demonstrated even more when the 2nd Battalion, 157th PRCT was ordered on D+1 to relieve the Regimental Reserve commanded by Captain Fraser and also to extricate the 1st Battalion Commander and his group in the vicinity of Les Arcs.

Meantime, the 3rd Battalion, 517th PRCT arrived at the 517th PRCT Command Post at Chateau Saint Roseline about 1600 hours on D+1. Despite exhaustion from the forced march the Regimental Commander, Colonel Rupert Graves, ordered his 3rd Battalion to take Les Arcs by nightfall. This action was essential as General Lucien Truscott, Commanding General VI Army Corps, wanted Les Arcs cleared of enemy by the following day. At daylight, the 3rd Battalion was 800 yards short of Les Arcs. The advance was resumed and Les Arcs was taken.

The town of Le Muy, which had been assigned to the British 2nd Independent Parachute Brigade Group as a D-Day Objective, proved rather difficult to take even though the British troopers had captured the main bridge leading into town well ahead of schedule. General Fredrick, Commanding General of the 1st Airborne Task Force, reassigned this mission to the newly arrived 550th glider Infantry Battalion which captured Le Muy on D+1.

The 551st Parachute Infantry Battalion with the 602nd Pack Artillery Battalion (Glider) attached, was ordered by the 1st ABTF at 1100 hours on D+1 to attack Draguigan and seize the town. Throughout that afternoon and into the night the 551st fought its way into this town. The Commanding General of the German Army LXII Corps, General Neuling and his staff, along with several hundred troops that included a special officer cadet class, surrendered. This was done in sufficient time to permit a special mobile force from the U.S. Army VI Corps to pass through the town on D+3 on their way to the Rhone and beyond.

The 509th parachute Infantry battalion was of great assistance to the amphibious forces by making early contact with these forces and subsequently easing their movement inland. Also of note is the fact that the 11 out of 12 pack howitzers of the 463rd Parachute Artillery Battalion were operative in less than an hour after landing. Likewise, the 4.2 Mortar Companies and the 602nd Pack Artillery Battalion (glider) which came in by glider were ready for action shortly after landing. The rigorous parachute training for the artillery units paid off and was exemplified by the 460th Parachute Artillery Battalion in the way they man-handled their howitzers in order to keep up with the ever changing situation that characterized this operation. Although insufficient targets materialized to require much use of this fire-power, the artillery units never the less were prepared to furnish artillery support when and as needed. It was comforting for the infantry troopers to know that they were there and ready.

By D+3 the 1st ABTF had commenced to reorganize in the vicinity of Le Muy. Following the reorganization it proceeded to advance along the Riviera towards Cannes, Nice and the Italian Border. The British 2nd Independent Parachute Brigade Group was taken out of action and preparations were made to return it to its base in the Rome area for further operational use. The First special Service Force replaced the British Paratroopers, and the 1st ABTF then continued to advance along the coast, meeting determined enemy rear guard opposition. These operations of the 1st ABTF toward the Franco-Italian border were not restricted to the coast, but extended to a point some 65 miles inland. As has always been the case when airborne troops are retained in the line in an offensive capacity, they experienced back-breaking difficulty in transporting their heavy supporting arms and ammunition. The fluid, rapid advance of the Seventh Army as a whole made it difficult for the Seventh Army to provide the necessary vehicles for the 1st ABTF. As a result, the paratroopers in many cases hauled their 75mm pack howitzers for some 60 or 70 miles over the rugged Riviera coastline. Fortunately, a number of captured enemy vehicles, together with the units organic transport brought in by gliders, did make the movement feasible.

While the 517th PRCT was used to portray the versatility and aggressiveness of airborne units the very same qualities were found and able demonstrated time after time by the other 1st ABTF units in DRAGOON. The 1st ABTF, at the close of the airborne operations phase of DRAGOON had all the necessary credits to recommend its retention as a Separate Airborne Force, to be retained at the Theater level, for repetitive airborne operations.

For all practical purposes, the airborne phase of DRAGOON was over by D+2 due to the earlier than expected link-up with the Amphibious Forces late on D+1. Combat actions continued in the airborne objective area due to the scattered nature of the combat actions, but the enemy was in complete disarray and had difficulty in conducting any coordinated actions of significance. From this point on the 1st ABTF was to play the role of a light special type of divisional size force as it protected the Franco-Italian border from enemy inroads. It was at point in time that the lightness of airborne units that made them so well suited for airborne assaults now went against them due to their lack of ground mobility which was the key to staying power in extended ground combat operations. It was this overpowering lack of mobility that caused General Devers, Commanding General of Sixth Army Group, to relegate the 1st ABTF to a security type of role.

Statistics

During OPERATION DRAGOON the Provisional Troop Carrier Air Division flew 987 sorties and carried 9,000 airborne troops, 221 jeeps and 213 artillery pieces. The sorties flown also included 407 towed gliders and carried over 2 million pounds of equipment into the battle area for the 1st ABTF. One aircraft was lost as a result of the OPERATION itself and the losses in aircraft from the period of movement from the United Kingdom to the conclusion of the OPERATION totaled only nine. No Troop Carrier personnel other than glider pilots were known to have been killed; 4 were listed as missing, and 16 were hospitalized. The balance of the 746 glider pilots dispatched on the operation had returned to their organizations.

On the airborne side of the picture, 873 American Airborne personnel were listed as killed, captured, or missing in action with 327 hospitalized on D+2. By August 20th, (D+5), this figure had fallen to 434 still listed as killed, captured or missing in action while many of the hospital cases had returned to their units for further action. The British 2nd Independent Parachute Brigade Group had 181 troopers listed as missing in action and 130 men hospitalized. Later reports indicated that 52 British paratroopers had been killed and that 500 replacements had been requested by the American units and 126 replacements from the British unit. By D+2 over 1000 prisoners had been taken by the American Forces and nearly 350 by the British Brigade. By D+8, this figure was well over 2000. The total jump and glider crash accidents amounted to 283 or approximately 3 per cent of the operational personnel involved.

The recovery of the parachutes both of the personnel and cargo types was very low. As of September 1st, not more than 1000 parachutes had been sent to the Rome base for salvage and repair. Similarly, the gliders that could be used again were very small in number. A survey of the LZs indicated that fewer than 50 of the gliders could be salvaged without excessive cost.

Conclusions, Summary and Final Conclusions

The following <u>Conclusion</u>, <u>Summary</u> and <u>Final Conclusions</u> are taken directly from a declassified Allied Forces Headquarters report on <u>Airborne Operation Dragoon</u> dated

September 4, 1944, that was prepared by the G-3 Section Airborne Advisor, Major Patrick D. Mulcahy. As this analysis unfolds, one must remember the information contained in this report is a snapshot in time immediately after the operation and a retrospective analysis. It should be kept in mind that this analysis was conducted by the Allied Forces Headquarters Report written by a member of the "airborne fraternity" who like so many of the airborne troops found the new from of airborne warfare as daring and exciting and did not understand the meaning of the words "it can't be done". The introductory paragraph of the Allied Forces Headquarters Report (AFHQ), under the heading "conclusions" is typical of this attitude. What follows is the Conclusions, Summary and Final Conclusions of the AFHQ Report:

Conclusions

1. The most obvious criticism of the airborne plan for "DRAGOON" was that it was not bold enough. The execution of the missions assigned the force were handily executed with precision. Although it is the worst kind of Monday morning quarterbacking, it could be said that a plan for a drop as far back as Grenoble could have been used to far better advantage, such, in retrospect, would have been better. A second point along this same line is the fact that instead of being used in the pre-H hour assault, the airborne troops could have been held in readiness until the withdrawal conditions became obvious, at which time they could have been dropped by D+2 or D+3 further inland to prevent any further retreat en masse up the Rhone Valley. The French Parachute Regiment is and has been available, and could have been used in this mission along with elements of at least part of the Airborne Task Force.

2. The airborne plan was well planned and thought through, and within the framework of the plan could hardly have been better laid on by the detailed planners.

3. The use of Pathfinder Aircraft and airborne Pathfinder Teams, above everything else, was well planned and executed. <u>Without the use of radar, radio and other aids, there simply could have been no airborne mission.</u> The development of the PPI used in the Pathfinder aircraft has made possible airborne operations which could not have been considered a year ago.

4. The routes selected for the mission were excellent. Utilizing all possible check points, shortest distances, avoiding dog-legs or enemy flak positions while making landfall by the prominent, irregular coastline—all while managing to stay out of the Naval channel in the prescribed 10 mile corridor, made possible the excellent results obtained.

5. Despite every precaution certain Troop Carrier A/C did pass over individual naval vessels. The Navy by their disciplined action demonstrated that the bitter lessons learned at Sicily had produced results.

6. The 334th Quartermaster Depot Company which is the American Theater Re-supply organization turned in a magnificent job in their organizing the supply base at Galera, securing the supplies and packaging them in time for the operation. This unit packed over 10,000 parachutes in a seven day period and prepared for dropping

five complete days of supply (sic) for the Airborne Task Force, a total of some 1000 tons.

7. The altitudes flown, speeds adopted, formation used both for the parachute dropping and air landing were excellent, with the exception that the "V of Vs" formation flown at an excessively high altitude for the supply dropping caused considerable drifting and scattering of supplies. Had these missions been flown by single aircraft in extended column at a lower altitude the necessary pin point dropping could have been effected (sic). This is particularly true because of the lack of enemy air and flak opposition.

8. Pre-operational training was excellent considering the short time available for such training. The airborne troops could have used one mass drop per unit on a final exercise but the time element precluded it. The combined Airborne-Naval exercise proved to be particularly instructive.

9. The choice and organization of the take-off airfields was good. Much credit must go to the 51st Troop Carrier Wing which made possible the rapid preparation and occupation of these airfields by much work prior to the se up; of the Provisional Troop Carrier air division (PTCAD).

10. Credit must also be extended to the Airborne Training Center without whose help the Provisional airborne Division could hardly have been effected. Despite the fact that the Center itself moved into the Rome Airfield Base Area but a few days before the Division (1st ABTF) they aided General Fredricks in countless ways to expedite his work.

11. the efforts made by the Airborne and Troop Carrier headquarters to have the Army, Nay and other air forces become familiar with their routes, procedures, formations and even individual troops were very successful. Squads of parachutists were taken in the various assault battalions of the 3rd, 36th and 45th Divisions in order to preclude any such mishaps as occurred on previous airborne operations when our troops fired on our paratroopers. The marking system employed by the Troop Carrier Command is also commended. The black and white stripes became well known to the Navy and other Air Force units prior to the operation and served to decrease any possibility of improper identification.

12. The glider assembly program was particularly well managed. By D-Day 407 well put together gliders were assembled and ready for operations, which was a brilliant feat since the vast majority of these gliders had but recently arrived form the United States by boat.

13. The weather forecasting while generally good, proved weak with respect to the direction of the prevailing winds during the flight. Had this wind been stronger or had the over water beacons failed, the navigators in all probability could have not made their off-sets which would have certainly jeopardized the complete mission.

14. The diversionary effort using the parachute dummies and "window' proved quite effective. It is recommended that further use be made of this equipment in future operations.

While the photo coverage of the area was generally good, most of the photos which were effective arrived at such a late date that any changes in the plans could hardly be made for fear of compromising the entire mission. This was in particular demonstrated by the late discovery of the anti-airborne poles on the DZs. Likewise more terrain models of a scale of 1:10000 or 1:25000 must be made available both to the airborne and troop carrier units for airborne operations. Copies of the small scale models used for the beach landings were not adequate.

16. Provisions must be made to bring in a service group with each airborne group landed. Parachute Maintenance recovery crews will have to be included in further operations unless we are to continue our extravagant use of parachutes. Such details must be given no other job than that of recovering and guarding parachutes. Similarly, the glider pilots after their landing has been completed could with a small crew act as guards for the LZs and protect the valuable gliders. In Southern France no guards could be found on the LZs and yet thousands of dollars of expensive instruments were in each glider.

17. A similar recommendation holds true for the air supply dropping. Crew members of the Air Re-Supply units should drop with the bundles to help their immediate recovery. Further, the Air Corps should not attempt to fly in re-supply missions in large unwieldy formation at high altitudes. Such missions require great accuracy if the supplies are to reach the proper personnel on the ground.

Summary

1. The recently completed airborne operation "DRAGOON" without doubt was the most successful airborne operation yet undertaken by Allied Forces. (See author's comments0. The Commanders of both the Airborne and Troop Carrier units are to be highly praised for the excellent manner in which they executed this mission. It is further hoped that the results and experiences of this OPERATION can be transmitted to all Theaters so that even those few mistakes made in this operation will not be repeated again.

Sources Used in This Account of DRAGOON

The primary sources for this post-WWII overview of OPERATION DRAGOON came from five principle sources. The Preface and Background sections of this report came from my own personal recollections, teachings, interest and research which have spanned more than twenty-seven years of involvement in the Airborne. The reminder of the material came from the following documents:

1 The 1st Airborne Task Force after action report entitled, "Report on Airborne operations in Dragoon", dated 25 October, 1944

2 The G-3 Section report on "Airborne operations Dragoon", dated September 4, 1944

3   "The Airborne Missions in the Mediterranean, 1942-1945, USAF Historical Studies No. 74" prepared by the Military Affairs/Aerospace Historian, Eisenhower Hall, Kansas State university, Manhattan, Kansas, September 1955

4   "Paratroopers Odyssey- A History of the 517th Parachute Combat Team", published in November 1985.

"The 1st Airborne Task Force Report", was used almost in its entirety. The Allied Forces Headquarters Report, which was based on the "feeder report' submitted to higher headquarters by the 1st ABTF was used as a back-up and 'filler' until the extracts from the AFHQ Report was used under the "Conclusions' portion of this overview under the original AFHQ headings of "Conclusions, Summary, and Final Conclusions". These three AFHQ sections were directly quoted from the final three sections contained in the AFHQ, G-3 Report.

"The Airborne Missions in the Mediterranean 1942-1945, USAF Historical Studies, No. 74", was used sparingly because of the mass of detail contained in the Report, which was difficult to extract and have fit in with the information contained in the "After Action Reports". "The Airborne Missions in the Mediterranean, 1942-1945, USAF Historical Studies No. 74" presents an excellent description of the OPERATION and gives a more complete detailed account of OPERATION DRAGOON. It was written some eleven years after the events occurred. "Paratroopers Odyssey- A History of the 517th Parachute Combat Team" was the main basis for the brief account of unit actions and thus the 517th PRCT, the largest of all the troop units participating in DRAGOON, was used for illustrative purposes for the airborne phase of this operation.

Bud was in Civitavecchia, Italy, preparing for the invasion. On the night before the invasion Bud decided he would get with one of the other paratroopers and take pictures of each other in their combat uniform with all of the equipment. These pictures have now become world renowned and appeared in Battling Buzzards, and First Allied Task Force books. Bud told the following story behind those pictures.

"I wrote the following on October 3, 1993. MY REMEMBERANCE OF THE TWO PICTURES THAT WERE TAKEN OF ME JUST HOURS BEFORE JUMPING INTO SOUTHERN FRANCE BEHIND ENEMY GERMAN LINES IN THE DRAGUIGNAN AREA OF FRANCE, IN "OPERATION DRAGOON." AUGUST 15, 1944, FROM THE MEMORY OF HARLAND LOREL "BUD" CURTIS

"These two pictures were taken at the spur of the moment rather than planned.

My friend from Headquarters, First Battalion, communications section, who I think was Joe Sumptner, from Kokomo, Indiana, took these of me, and I took some of him. When we decided to take these pictures we used my camera and Joe's equipment. The first picture is me standing at the tail of a C-47 airplane. I am dressed as I was on the morning I jumped except Joe's trench knife is shown on my right side below the canteen. Some individuals thought I was wearing three knives, but what appears as a knife scabbard on my right waist is a bayonet scabbard. I packed the film with me for about three months before getting prints made in Nice, France.

The preceding events leading up to the time, place, and circumstances of taking these pictures began after a two-week ocean voyage from Newport News, Virginia, to Naples, Italy. This was in the last half of May 1944. We were bivouacked for two weeks south of Naples in a place called Bagnoli. Not too much farther south was Pompeii. I did see both places before departing that area via LCI's (Landing Craft Infantry). When we left the USA, we were destined for Anzio, which had been a terrible place of carnage. But Anzio fell before we arrived. Rome fell June 5, 1944, and D-Day was on June 6th, the invasion of

Normandy, France. The LCI's transported our 517th Regimental Combat Team to about fifty miles north of Rome. I remember places like Grosseto and Fellonica.

We got our first introduction to real war against the Germans in this sector for about six weeks in June and July of 1944. Around the middle of July the 517th was trucked to a place not too far south of Rome. I recall a place called Franscatti was within a walking distance of five miles or so. There was a big lake there that was actually in the bottom of a volcano that had been extinct for hundreds of years. It was a good place to go when possible, as it was very hot and humid around Rome in the summertime.

Later, we learned the reason we were there was for intensive training preparatory to the invasion of Southern France targeted for August 15, 1944. About three days before D-Day we were transported to an airfield outside Rome in Civitavecchia. Tight security was enforced. Much time was spent at this airfield getting equipment ready plus studying maps and sand tables showing where we were suppose to land, which was behind Marseilles not far from Grasse, the perfume capital of Southern France. The invasion was code named "Dragoon". The map and sand table studies were a waste. The Air Force (Army Air Corps) jumped our troops in at least a fifty mile radius even into the Mediterranean sea where some were drowned. I landed in tree-thick mountains (LesArc area) far from the intended drop zone that turned out to be a stroke of fortune. The few who did land on the drop zone were annihilated. Berlin Sally announced on the radio August 14, 1944, that they would have a hot welcome waiting for the 517th. They had gasoline-soaked haystacks and 88 millimeter shells attached to piano wire strung from eight-foot poles which proved to be as hot as she said, and totally lethal for those who did land there.

Now back to Civitavecchia. It was during a brief period on August 14, 1944, that I got the idea to do the pictures using one of the C-47's as a prop. The results were far more than I expected. Just having these pictures today is a minor miracle when thinking back what the negatives and prints went through to be in this picture frame today is remarkable.

The pictures now appear in Gerald Astor's book, "Battling Buzzards" and Michel De Trez's book, "First Airborne Task Force", pages 152 and 153. Both books provide an account of the combat experiences of the 517th Parachute Regimental Combat Team (PRCT). The experienced eye will see that I am wearing combat boots (Army nomenclature known as boots, service, combat, composition sole, M-1943). These boots laced up in the lower half in a conventional manner with a wide two-buckle cuff at the top.

I left my shined-up paratrooper boots behind (Army nomenclature known as boots, jumper, parachute, M-1942). The paratrooper boots were the envy of every ground soldier since only the men of elite paratrooper units were authorized to wear them. I wanted them in good shape in case I did return. Another thing noticeable are the two trench knives. The one on my leg is mine. The other one on the harness belonged to my friend Joe Sumptner. The folding-stock M-1 carbine rifle was not in the case that accounts for the empty case appearance. These pictures were taken about 4:00 P.M. on the 14th of August 1944, just about twelve hours before the actual beginning of the invasion of southern France that became the real thing at 4:15 A.M. on August 15, 1944. The rest is history. Perhaps someday I can tell some of the many side stories that are associated with the above main story."

Pictures taken by Bud's friend with his camera on August 14, 1944

This is the camera Bud used to take the famous pictures on August 14, 1944.

After Bud took the pictures of his friend and of himself he put the camera into his mussett bag. This bag can be seen in both pictures just under his reserve parachute. The picture of Bud standing in the woods was taken a few days after his combat jump into southern France, near Le Muy

On the next page is the TOP SECRET letter written by LTC Graves, and was read to each man before he jumped in Southern France, 15 August 1944.

TOP SECRET
HEADQUARTERS 517ᵀᴴ PARACHUTE INFANTRY CT
APO 758, U. S. ARMY

11 August 1944

1. The following general considerations should be noted in the operation now being undertaken.

    a. For most of us this is our first combat jump. Consequently some may be a little apprehensive. Remember that the advantage is with the attacker, as the enemy does not know exactly how or when he is going to be struck. Particularly in an airborne operation in which we land in his rear areas where his CP's, lines of communication and supply echelons are set-up, will our activities give him grave concern. The enemy consequently will be a lot more apprehensive than we are. That the enemy will react to our landing by movement of forces towards our area should be expected within a few hours. However by that time we expect to be pretty well set for him and deal out a lot of punishment. It must be remembered also that he will be engaged at many other points by other airborne units and the large scale attack by the amphibious landing. There can be no doubt of the success of this operation if we use our heads and keep our confidence, work quickly but smoothly and act aggressively using good tactics and security measures.

2. As it will be dark when we first land it will be difficult to see what is going on around us. Therefore, we must not start firing promiscuously at any thing that moves. You must be certain it is the enemy before you fire. After the first unit lands many of our men will be moving around the area, some already in assembly or moving to assembly positions and others recovering bundles. Don't fire first and find it is one of your own men later. Weapons, on landing, should be loaded and locked, and fired only on orders of an officer or in case of emergency. The sound of enemy weapons is known to you and should disclose the location of enemy forces if present. Enemy flares may be fired around the area to give the appearance of enemy strength and to cause us to be alarmed. Remember that more casualties can be caused by some of our own men getting trigger happy than from enemy fire. It is possible that your stick may land some distance from the DZ. Your action in this case should be considered. The general idea is to move to your Battalion assembly area and if the Battalion has already left for its objective join them there. If this is impossible join up with friendly troops preferably of this unit and assist them in accomplishing their mission. In either case do as much damage as possible on the way, for example a staff car or a truck load of Germans may be driving along the road unaware of your presence in the area. In case no officers are present with your group the senior NCO should be prepared to take over. All men will be issued instructions regarding the terrain objectives, important towns, streams etc., and also a map in the escape kit. Remember that a few men can create a hell of a lot of trouble if they happen to be in the right place.

3. Due to lack of transport in landing and the difficulties of supply, don't expect any easy time. However, I am confident that this outfit can take care of itself in any situation that may arise. May success reward our efforts, and good hunting to each and every one of you.

R.D. Graves,
Lt. Col., 517ᵗʰ Prct. Inf
Commanding

Letter to Mom from Harland "Bud" Curtis

Combat Jump into Southern France
August 15, 1944 as recorded by Bud on August 22, 1944

Dear Mom,

We boarded C-47's (the twin engine plane that was used by paratroopers) in Italy (Chiteviccia) about 2:30 am and had a nice pleasant ride with no opposition at all. Most of us were asleep until almost time to jump. They woke us up and said we would be over the field in eight minutes. That was about 5:00 am. We stood up and hooked up. It seemed like years went by as those last minutes ticked off. I was number 13 man. The green light came on and guys began to disappear in front of me. Then there I was at the door. I had a hell of a body position. I went out of the door like I was throwing a flying block with my right shoulder at somebody. I was heading down nose first when "Wham" she opened and jerked me back up right. I looked up to make sure my chute was open and then I looked around. We must have jumped awfully high because I thought I was never going to come down. There was a low fog about a 100 feet off the ground and it looked just like water. I really thought my number was up for sure. I was cussing the Air Corps and all their ancestors for 17 generations back.

When I sank through the mist I was just beginning to figure it all out when "Thud" I hit the ground. I will never forget that morning. I was miles away from the jump field. Later I found out that it was a good thing I didn't land on the jump field as the Germans had it all ready for us with mines, machine guns, and flame throwers. All I could see was forms of trees through the fog. I cut myself out of my chute and when I stood up I seemed to have lost my sense of balance. I fell down and rolled down the side of a mountain a few yards. I stood up again, and did the same thing again. I stood up again and took a couple of steps and fell off a ledge about 10 feet high and about broke my neck. There was dry grass all over and every step I took you could hear it for a mile. I decided to lay still for a while and see if I could figure out where I was at. I didn't know which way to go. I heard somebody moving a little ways in front of me. I shouted the password at him hoping it was one of our guys, but instead of getting the right answer I got a couple of bullets just over my head. I took off for a big rock and figured I would have it out with the guy, but then I heard somebody behind me. Once again I made the mistake of hoping it was one of our guys and shouted the password to him and got my answer in hot lead. It was so foggy we couldn't see each other but we could hear every move each of us made.

There must have been a whale of a patrol around me and every step I took away from them I could hear them coming closer. I knew as long as it stayed foggy I could hold them off, but it began to get light and I decided the best thing to do was make a run for it and hope they would miss. I took off zig zagging and they opened up on me, but I was lucky and got to the other side of the hill and down in the valley and there I met some of our own guys.

We climbed over another hill and came to a road and met up with most of the company. Ever since then I haven't had much trouble. In fact the Germans are running to fast. I haven't seen one for days.

End of the letter

In Bud's letter dated August 22, 1944 he described that he was the thirteenth man in the stick in the combat jump. Many men may have thought being the thirteenth man to make a combat jump to be unlucky, but not Bud. He was born on the 13th of December, and later after the war was married on the 13th of June. Thirteen had always been his lucky number, and still is today.

In 2004 an article appeared in the Thunderbolt newspaper (a newspaper started for the 517 PRCT during the war and continued after with the 517th PRCT Association for its members). Bud had read an article about a challenge and password used on that first day after their combat jump. The password did not match with what he remembered so he related to me what he remembered about his jump on August 15, 1944. He asked me to write in to the 517th email site "Mail Call." This site is used daily by members of the 517th, their family and friends. Bud's comments started a great deal of conversation on the email site from many troopers and what they remembered about the challenge and password that day. Some men had clickers like the 101st Airborne Division did on D-Day in Normandy. Like my Dad some did not have those clickers.

This is from "Mail Call" the 517th email site:

More about the challenge and password.

My father (Bud Curtis, HQ, 1st BN) did not have a clicker, but was only given the challenge and password. He told me last night 2-11-04, when he jumped into southern France all of the information he was given didn't match. The moon was supposed to be on his right side, with trees on the left. He said it wasn't. All he saw was clouds below. With the moon shinning on the clouds my dad thought he was in water. He began to try and release himself from his harness so he wouldn't drown. The harness was too tight and he couldn't get out of it (thank goodness for him and me!). When he landed, it was on a terrace. He still couldn't get out of his harness and had to cut himself free with his knife. He said it was pitch black and he couldn't even see his hand in front of his face. He said he must have landed right into a German patrol of 5 or more men. He said he took a couple of steps and fell down the side of the terrace to a lower terrace below next to a tree. There were leaves on the ground and every time he took a step it made a crunch sound. He heard someone and so he quietly said, "Democracy", the challenge. Then waited for the password, Lafayette. It never came. Just then he could see the silhouette of a German soldier though the haze, about 100 feet in front of him. He also heard other Germans off in the distance to his right crunching on the leaves. The German to his front raised his rifle. In that split second my dad was going to shoot him, but remembered his gunpowder flashed and the German's gunpowder did not. He knew if he shot the German to his front, the Germans to his flank would shoot him when they saw the flash from his rifle. He didn't shoot but dove to the ground just as the German to his front shot. My dad heard the bullet wiz by. He laid there quietly until the Germans walked off in the other direction, I guess thinking they killed him because they heard him fall. He lied there for a few minutes and then got up and took off in the opposite direction and heard someone else. He said "Democracy" quietly, no answer. He said it louder, DEMOCRACY! He next heard an American voice say, "Shut up your going to get us killed." He shut up and was very thankful to find another American Paratrooper, of which to this day he never knew his name, but they sure took care of each other on that day. Bud also said that if a trooper forgot the challenge and password he was to say, Billy the Kid.

Bud remembered shortly after the jump on that first day, one of the men in his communications section was ordered to climb a telephone pole and cut the lines. The trooper didn't want to do it, but he followed the orders. As he attempted to cut the power lines he was electrocuted, and fell dead on the ground. That same day one of the troopers ran up to a German fuel dump and threw a grenade to destroy the fuel. As he the grenade exploded he caught on fire. Bud described this trooper as being badly burned, but the men were not able to stop its mission so this trooper had to continue moving with them. Bud described this trooper as being in constant pain, but noting could be done. As the men stopped to rest for the night this trooper was then taken to an aide station. Bud never knew what happened to this man after that night. He may have lived or he may have died.

That first day in France was a rough one for Bud and the other troopers. As Bud tried to sleep on the ground, and put out of his mind the horrible things he had seen, he heard the voices of an officer telling two sergeants to take some German prisoners back to the aide station. A few moments later Bud heard the Germans yelling Nine, Nine, Nix, Nix! Then gun shots. Bud then heard the sergeants return, and the officer asked what had happened? One sergeant said, "They tried to escape." Bud has never forgotten that night.

Little time was left for Bud to write letters after his combat jump into southern France. On August 20, 1944, the 517th moved north from Puget-sur-Argens to participate in the "Champagne Campaign." They had no vehicles so they marched. They had no water and so they obtained water from nearby streams, and purified it with halazone tablets. With no blankets, and only the clothes on their backs they had from the combat jump, they marched up and down the mountains of the Maritime Alps. Just north of the Argens valley the terrain rose up and then down with deep gorges going from northwest to southwest, cutting directly across the Combat Team's route. German resistance was relatively light, and was encountered in only a few locations, but men still died. In August it was hot in the daytime and the men were drenched with sweat. At night the temperatures dropped, the men were cold and shivered all night. They were clothed lightly, and had been wearing the same boots since leaving the United States. Their boots and clothing were wearing out, but there was no resupply so they had to do with what they had. The entire regiment was unbathed and unshaved.

The mission of the 517th PRCT was to protect the Army's eastern flank. This meant LTC Graves had to move his regiment as far east as practicable and then defend the best ground available. The initial Task Force objective was the line Fayence-La Napoule. The 517th was assigned to the left flank, the British Special Service Force had the center, and the 509th and 551st had the right flank. The German 148th Reserve Infantry Division planned to make a fighting withdrawal to the east. The Germans did not want to fight in Nice because of French resistance and crowed streets. Instead they made their stand along the Fayence-Callian ridgeline and the Siagne and Loup Rivers. The 517th had the mission to seize the line at Fayence-Callian. This they accomplished and were given the task of seizing St Cezaire which they did by the 23rd of August 1944. The Germans defended St Cezaire and put up a fierce fight there. By the time the fighting was over twenty one Germans were dead, and five Americans with numerous wounded on both sides. The enemy's determination to hold St Cezaire was evidenced by the mines and booby traps found by the 517th Engineers.

Bud and the First Battalion moved to St Vallier and on the 24th of August 1944 the 517th occupied an area north of Grasse. Another day of hard marching brought the unit to Coursegoules, Bezaudun and Le Broc, areas 10 to 12 miles northeast of Grasse. On August 26th, 1944, the First Battalion moved to Carros near Var where free French Fighters were suppose to be surrounded by the Germans. When the First Battalion got to Carros there were no Germans. The First Battalion had to cross over a bridge at the Var River.

Only two German sentries were there. They were quickly killed and the men moved into Var. East of Var the Maritime Alps rose steeply from the Mediterranean. According to Archer (1985), "The zone of the 517th can be visualized as a large triangle with the apex at the crest of the Authion Massif peak, just north of Turini, Italy." This peak was 6,380 feet. The west leg of the triangle is the Vesuble River, which runs deep through gorges to the Var river near Leven. On the east is the Bevera River, which flows south from Authion to Sospel and turns east to the Italian frontier. At the base of the triangle is L'Escarene, where roads from the west and south converge to go north to Turini, Italy and northeast to Sospel. The mountain ranges formed strong defensive barriers against advancing American troops from the west. Tete de La Lavina, controlled the approached to Col de Braus, L'Escarene, and the Sospel road which leads to Italy. The Germans withdrew and moved into these mountains. The First Battalion moved into the Vesuble Valley, occupied St Jean-la-Riviere, and moved toward Lantosque. Stiff German resistance appeared on August 31, 1944, when a motorized patrol from the First Battalion was ambushed east of La Bollene. Three men were killed, two wounded and a jeep destroyed. Bud writes to his mother on August 29th 1944, and tells her he hopes he can make it home. He does not tell her about any of the fighting that has occurred.

Letter to Mom from Harland "Bud" Curtis
Letter written on the back of a letter Bud received from his mother in pencil
French Alps, Southern France
Monday, August 29, 1944

Dear Mom and Family,

I haven't got any stationary so I am using your letter I got yesterday to write on. Stationary and envelops are kind of scarce up here in the French Alps. I traded half a can of my only rationed pineapple juice for 4 envelops. These guys who got packages are making me envious so I am going to ask for more things. From what other fellows got, I think caramels, butterscotch, peanut brittle and cookies hold out pretty good in the mail. I would like to have a can of honey, fruit juice in fact I would like anything you can send and don't worry about what condition it arrives in as you can be sure I'll eat it. I can use some stationary as you can see. I wrote Jill a long letter today, but I don't like to ask her to send me anything or you either. I guess you are kept pretty busy without me having you sending packages. Just keep the letters and pictures coming and I'll be happy. How about my watch? I could have good use for a pencil now. This one is about shot.

We have a short wave radio set up and I am beginning to find out what is going on in the world. It is beginning to look like we are getting them beat over here now if they get all the Japs out of the south pacific. I don't see how the war can last much longer. I hope I don't go to the Pacific. There is nothing there I want to see, in fact if I ever set foot on the U.S. again they will have a heck of a time getting me out again. I am going to lead a nice quiet peaceful life. I have had enough adventures to last a dozen people a life time, but I guess I will have a few more before I finally settle down to that peaceful life. I hope I don't run my luck too far. Stepping out of an airplane over a foreign country into the middle of a German patrol isn't my idea of a peaceful life. I don't ever want to do that again (Bud never did make another parachute jump. His combat jump on August 15, 1944 was his last jump ever). *It makes you wonder if you are ever going to see good ole California again. I'll tell you about that deal sometime. Have you got the money I sent home and that $45.00 allotment? It will be payday again soon so I will be sending some more and another $45.00 also. That is every month so let me know if you get them all.*

I hope you give Jill something from me on her birthday. Wish I could be there.

Write soon.   All my love to all,   Bud

Bud had written on the backside of a letter he received from his mother. It is the only record of his mother writing to him during the war. The many letters Bud wrote to his mother were answers by the many letters she wrote him. The letters she wrote to Bud while he trained in the United States and in combat overseas were destroyed by Bud. He could not take a chance of these letters falling into enemy hands. If captured they would surely be used against him. This is the only letter from his mother in existence. The only reason it exists today is because Bud did not have any paper, and he had to write his letter on the back of his mother's letter. Bud crossed out his mother's words on the other side of the paper with pencil. Fortunately, the letter could be read through the pencil marks. This is what his mother wrote to him from Long Beach, California.

The only Letter left in existence from Bud's Mother

Long Beach, California
August 9, 1944
(Wednesday)

Dearest Buddy,

Hurray! The mailman just left two V's (V mail letter) and an airmail. The V's were July 21 – 26 (1944), the airmail 29th (July 1944). Am I the lucky mother. It helps so much, with Bert having to sign up again today I've just been so blue for we hadn't heard from you for nearly two weeks. Thanks dear for these wonderful letters.

So the little package got there okay. I'll send more. No requests for them, but for the five pound boxes. Those little ones are packed and sealed so I really didn't know whether the stuff in it was any good or not, so just sent it as a trial, now you tell me it arrived alright and tasted good. You're going to get more, if Thrifty (she crossed out Thrifty store and wrote) Sontag still has them, that's where I got that one. I wish now I'd sent more of them, but I was afraid it might be soggy stuff and now I know, you can look for them often. The 5 pound boxes I've sent have some home made junk in them and I hope it gets there okay and tastes alright. Let me know what travels best and tastes best. I just talked to Jill on the phone she was at Walkers she said she mailed you six pictures last night and has the glamorous one out of hock and is going to send it. She has been trying to get a frame, they are scarce, so I told her to forget it and send it the way I sent the other one, so I guess she will. Did you ever buy those silk stockings or nylons? Oh my, what a thrill. We haven't seen any for nearly 2 years now. The aren't rationed. They just aren't any around so I imagine Jill will be mighty thrilled about the ones you sent her and I hope they get here for her birthday. I am glad you have a desire to save and be able to better yourself and the world when you get back. Only time will decide which you really need most. Education or a job. I imagine a good education means assurance of a more successful future, but if you stepped into the right job when you get home, you probably could do both with adult education. Personally I think it would be nice to come home and be young for a while again. Go to school, get a sheep skin, and a side job that would help you go into business for yourself. As I said, only time will tell. The letter ends here.

Letter to Mom from Harland "Bud" Curtis

Somewhere in Southern France
Sunday, August 27, 1944

Dear Mom,

Here it is D+12 – I have been in this country 13 days now. Right now I am way up in the French Maritime Alps. I have climbed more mountains than I ever want to see again. "Wow", I am even beginning to look like a mountain goat. We got our first mail today and was I ever glad to get one from you and two from Jill. This is the first time I have had time to write since I came hurtling down through space in the early hours of D-Day. What an experience that was, but I have been pretty lucky and haven't as much as a scratch on me. Three of my best buddies have been killed since we got here. I will have a lot of adventures to tell you when I get home that I can't write about now. I would give anything for a good meal and a month's sleep, but I am just thankful I am alive, and still have hopes of breezing up the front walk someday. I will write whenever I can, but don't worry if you don't hear from me. Tell Jill hello. I guess Bert has shipped out again. I hope not though. How is "Harland R." (Bud's new nephew, Bert's son) doing? I will sure be glad to see him. I would like to be home for Jill's birthday, I hope she gets those silk stockings I sent from Rome. Maybe I will be home for her next birthday. Notice the new A.P.O. (Army Post Office) 758. We are getting pretty well organized now so maybe I will be getting mail and I will be able to write more often. Send more candy or anything you can. I could really go for a package from home right now. I haven't got the other (packages) yet, but still have hopes. So long for a while - Lots of Love Bud   P.S. Send some Stationary

Letter to Mom from Harland "Bud" Curtis

French Alps, Southern France
Tuesday, August 29, 1944

Dear Mom,

  No letters today, but I got two big packages and was I ever happy. Everything was swell. The package was kind of beat up, but the things inside managed to survive. The pineapple really hit the spot with those cookies you made, I sure like them. The candy in that little tin box is the best I have tasted in many months. I really go for those caramels and fancy stuff. That little tin box sure did look familiar. I can remember you keeping odds and ends in it ever since I was big enough to reach the top drawer. I never knew I could appreciate anything so much. Home made cookies clear from California. What next? You can be sure I will keep asking you for things from now on. Send the same things again and add a little stationary. I traded some stamps for this stationary I am writing on. It is kind of scarce, but so are stamps and I have quite a few stamps so I guess I can make out for a while by trading. It has been a long time since I have eaten pine nuts. They sure taste good. Keep'em coming.

  We are still taking life easy. Three days now and has it ever been nice just sitting around resting up. We really had a work out the first twelve days. We climbed mountains that a goat couldn't make. I don't know how long we will continue taking it easy here, but it is nice while it lasts, and you can imagine how happy we all were to see the supplies finally catch up to us with mail and packages from home and also be able to write again. I guess you were beginning to wonder what had happened to me! I will write every chance I get, but don't worry about me if you don't get letters from me regular, it will just be because I am some place where I can't write. I hope I get some mail pretty soon telling me if Bert has shipped out yet. Your latest letter was August 10, 1944. Hope that everything is running smoothly at home. How is Lorrain and little Harley (short for Harland)? I'm sure proud that he was named after me (Bud's first name is Harland). I guess Reid is too (Reid was Bud's cousin and Harley's middle name was Reid). Does Jill come out to the house much anymore? I like your letters where you write about her coming out for the weekend. Boy, but I will be glad to get home again. So will millions of other guys. I'll just have to wait my turn.

  Keep the letters and pictures coming and a package now and then. Thanks again for the package.

  Millions of love,
  Bud

Letter to Mom from Harland "Bud" Curtis

Southern France
August 31, 1944

Dear Mom,

  We crossed 16 miles more of these French Alps yesterday and today we have been taking it easy. There is a nice stream just down the mountain and places where it is deep enough to swim and dive. I had lots of fun and boy that cool water felt nice. We haven't had any mail come in for the last couple of days so I will be satisfied to read over the ones I have until we have another mail call. I sure wish I could have been there for Jill's birthday today, but I should surely be there for her next one. Write and tell me what she did and if *you gave her anything from me. I hope so. Thanks again for those two nice packages, I am still eating from them. How about some more cookies and also maybe some stationary. Ink and anything in cans that doesn't cost points. Do baked beans cost points?* (During the war food items were rationed back in the United States) *If not send me some, and some more pineapple too, and a can of honey. I don't know which things take points so if I should ask for something that does just forget it.*

Is Dad still working at both jobs? I'll bet he is sure tired by now. I'm still waiting for that letter telling me if Bert has shipped out again (Bert junior Bud's brother served in the Merchant Marines). *Give my love to Jill, Lorrain, Harley, Dad,* (Jill was Bud's girlfriend, Lorrain Bert's wife, Harley, Bert and Lorrain's son) *and everyone else. I'll be glad when I can see my nephew. How is he doing? What has ever happened to Bob Douglass?* (Bob was Bud's cousin who went into the Merchant Marines. Bob was the son of Ross Douglass. Ross was the brother of Hyrum Douglas who married Bud's aunt Mada) *I haven't heard about him for a year now or Bill either. I haven't heard from Dean Hildreath. Has he written his mother lately. See if you can find out his address for me.*

I guess that about covers everything for now so I'll write again when I can and I'll be looking for some letters from you and Jill next mail call.

Until then, all my love

Bud

P.S. Send a few more stamps

Bud was covering lots of ground moving up to the tops of the Maritime Alps into a little town called Peira Cava. He talks about marching 16 miles up the mountain. With all the stress of combat and marching in the mountains he still takes time to ask his mother how his brother, wife, nephew, and friends are doing back home in his August 31st letter. Four days pass before he writes again. The men are conducting patrols, counter-patrols, and outpost actions to see if there is much German resistance. Fortunately nothing much occurred. It was the next best thing to a rest camp and the 517th deserved the break however; a soldier could get killed on patrol very easily. The Germans had troops in the area that were mostly young poorly trained and inexperienced. They were no match for paratroopers. Bud and the First Battalion were assigned to Peira Cava. LTC Boyle led the Battalion up a hair pin road from Luceram. While there they captured a field order that contained information about where a German patrol would be. Men from the First Battalion, 517th went there and sure enough, there were the German soldiers. They were emplacing mines along the trail. The men attacked the German soldiers killing two with the rest scattering. The 517th figured the Germans would be back the next day so they went back to the location and set up an ambush. Sure enough the Germans showed up with shovel and picks to lay more mines. The ambush killed six Germans, captured five and the rest fled never to return.

While the First Battalion held Peria Cava, the rest of the regiment was engaged in a battle to take Col de Braus. On September 3, 1944, the 517th PRCT with Second and Third Battalions moved into position to take the town of "Col de Braus." Second Battalion initiated the attack on Col De Braus. The Germans did not want to give up the mountain town, and began a series of counterattacks, supported by mortar and artillery fire. On September 6, 1944, the regiment established a defensive position. The Germans were desperately trying to hold onto the territory, but more and more Americans were now in the area. The mountain Tele de la Lavina dominated the entire area. It was three-quarters of a mile southeast of Col de Braus, and Tele de la Lavina would have to be captured before Col de Braus could be secured. On September 9, 1944, an attack was launched to take the mountain, and the Germans let go with everything they had. The Germans fought at point blank range with grenades, machine pistols and "Panzerfaust" antitank projectiles. The battle continued on until the 517th was able to take Tete de la Lavina. Twelve days later on September 18, 1944, Tete de la Lavina exploded in a rain of steel and fire. The battle was fierce. Thirty to forty German soldiers were killed. Sixty one enemy soldiers surrendered. Col de Braus and the surrounding areas were finally secured, and in American hands. Over the next few days several German counterattacks were beaten off, and the Germans fell back to Sospel. Col de Braus was finally taken, but it had not come cheap or easy.

After the fighting stopped in the Piera Cava area, Bud wrote home on September 3, 1944. He expressed to his mother how important life was. He wants to be able to eventually come home and marry his girlfriend. He explained to his mother he had many close calls of being killed. This caused her to worry greatly about him. All she could do was hope and pray that he would come home from the war. In Bud's closing sentence of his letter he expressed optimism about surviving the war and coming home. It is a good attitude. It gave his mother hope.

Letter to Mom from Harland "Bud" Curtis

Southern France
Sunday, September 3, 1944

Dear Mom,

I just received your letter of August 10th answering all the questions I asked you about Jill. I don't know why I ever wrote you that letter anyway.   I hope I have everything cleared up now as for Jill I have all the faith and trust in her possible, and you can bet your life when I get home this next time and I have anything to say about it we are going to get married whether I am out of the Army or not. I'm through worrying about other things. I am more than happy just to still be alive and when I think of times, and probably more to come that I have about gotten myself killed it makes me appreciate it just that much more.

There are lots of things I don't tell you because you would only worry and anyway you people over there couldn't possibly ever realize how things are over here, but anyway I don't give a darn if I am not out of the Army when I get home next time we are either going to be married then or not at all. All I know is that I want her and I am darn tired of all this waiting so that is how it is. I'll be home you can be sure of that. I've come close enough so many times already of getting killed if they couldn't do it then they sure can't do it anytime. I'm planning on living for a long time yet so the guy who tries to change my plans had better watch out. Well I guess I have gotten things explained now so you know how I feel. Just keep letters, pictures and packages coming now and then and I'll be home before you know it.

All my love
Bud

Letter to Mom from Harland "Bud" Curtis

Tuesday, September 5, 1944
"Viva La France"

Dear Mom,

Just a short letter to let you know I am alright and still in the little village taking life easy. Your letter was of the 17th of August and you said you thought I was in France, well I guess you have gotten my letters now so you already know you were right.

Send a few more stamps in your next letter. I traded some for some stationary and envelopes, but I have enough to last me until you send some more. I have Bert's new address, but I am not going to write him because I just got a letter returned to me that I mailed him June 27, 1944, and it just said, "Not on ship." I know he leaves a forwarding address, but

I guess they just don't bother to send it on to him. I don't want to write if he's not going to get 'em so whenever you write him say hello for me too.

I'll sure be glad when I start getting mail answering the ones I have written since I have been here. What I don't like is when I get letters all mixed up. One at the end of the month, then one from the first of the month and sometimes I don't even know what you are talking about until I get some back mail which are the missing links and sometimes when I get your latest letters first the others sound like back news, but never the less I am pretty sure that I eventually get them all and that is what counts so keep 'em coming and I'll make out okay. If you want to send me something how about a can of jam and some crackers. Send me anything like that I wish I had some now.

It is getting dark so enough for now. I'll write again tomorrow.

Lots of Love Bud

Bud Curtis standing in another one of those long chow lines.
Bud with two friends somewhere in France, 1944.

Letter to Mom from Harland "Bud" Curtis

Friday, September 8, 1944
" Up Front in Southern France"

Dear Mom,

    I haven't much to write about. We have moved up front again today. I am O.K. so don't worry about me. I was talking to a guy in "C" Company a few minutes ago and he said Charles Daigh was out on patrol today and happened to look behind him and spotted 3 "Jerries." He killed two of them, and I guess the other one got away. That's about all the local news for now. Lots of Love Bud

P.S. Send me some of that candy you have on hand and say hello to Jill for me. I won't have time to write her today as it is almost dark.

Letter to Mom from Harland "Bud" Curtis

Southern France possibly in Peira Cava in French Alps
Wednesday September 13, 1944
Bud returned there in Oct 1969, and June 2004

Dear Mom,

    It is late at night and I am writing this by candle light. I am still in this village and have a nice warm house to sleep in as long as I am here. This house is where we have our switchboard and the people that own the place are very nice. A young married couple their *baby and mother-in-law. I can either heat up my food or their stove or just walk down the street about 100 yards to a big hotel and use their kitchen and eat off of plates and not even have to wash the dishes, but there is a cute little French girl who washes the dishes and so naturally, I help her dry some dishes every so often. I am learning a few words now too. That Italian language was easy and when we were at Rome I was just beginning to get along fine. French is much harder to learn, but I like the French people much better than the Italians. I got another package from you yesterday and you put in just the things I wanted.*

That pineapple is sure good and with the cocoa you sent all I have to do is walk out to the next room and take some hot water off the stove, they always have some, and there, I am quick as two shakes of a lamb's tail, "cocoa" steaming hot. At the rate I am using it up though you had better get some more on the way fast. I sure appreciate the things you send so keep'em coming please.

I spent my last franc yesterday and bought 10 rolls of film for my camera and I have the ones you sent me in the other package so now I have enough film to last me the next 6 months. I'll have a lot of pictures to show you of the places I have been when I get home. I could use some airmail stamps. I only have eight left so put some in your next letter please. Thanks.

The weather has been pretty bad today. Rain and a cloud has settled down right on top of us. We are right on top of a high mountain. It wouldn't take long to go blind writing by this candle so I am going to sign off now. I hope I get a letter from you and Jill tomorrow. I'll write again soon but had better close this now with all my love. Bud

P.S.

I got a letter from Jerry and answered it. Tell her to remember the picture next time and tell her thanks for writing (a girl who worked at Sears with Bud's mother. Bud never met her until he returned from overseas. She married Bud's friend Bob Pack).

On September 21, 1944, the First Battalion was ordered to report to Col de Braus to relieve the Third Battalion. The Third Battalion moved back to Peira Cava where Bud and the First Battalion were. Now Bud was in the thick of it. Traveling on the main supply road beyond Col de Braus was named "The Bowling Alley." Any time vehicles or men traveled this road they came under fire from Germans patrols. Artillery and mortar exchanges went on constantly. Bud related a story of Lieutenant Lynch driving him and other soldiers in a jeep down this road from Peria Cava to Col de Braus. Lieutenant Lynch arrived at a hillside, and stopped the jeep just behind a hill where the Germans could not see him. He told the soldiers to hike down the hillside to where one of the companies of the first battalion was located. Their mission was to repair phone lines. As Lieutenant Lynch drove away, he hollered to the men saying, "Watch out for mines." This of course made Bud's blood run cold as they traversed down the hillside. Fortunately, they did not step on any mines. Sometime between September 13 and 23, Bud and the communication section moved into the tunnel at Col des Braus. For the next five weeks he would be held up in the tunnels at Col de Braus.

Except for the relief at Peira Cava the 517th sector remained unchanged until October 28, 1944. On October 23, 1944, the Germans lobbed in thirty to fifty artillery rounds daily on Col des Braus, Hill 1098 receiving 200 or more rounds a day. Bud suffered this intense German barrage and speaks of it often in his letters. It was the firing of these shells that would affect Bud so greatly for the rest of his life. He would never forget the sound of incoming shells. In June 2004, Bud returned to visit the tunnels at Col de Braus, remembering those horrifying days. Still on the ground and hillsides sixty years later were fragments from German artillery shells. Bud writes about living in the tunnel and the German shelling that took place.

Letter to Mom from Harland "Bud" Curtis
40th Day in France now in the tunnel of Col de Braus
Saturday September 23, 1944

Dear Mom,

I haven't had any mail for about a week now, but it seems that not much mail had been coming in, so there must be something haywire along the way somewhere. I sure hope someone gets it straightened out soon and I begin getting mail again.

I have left that nice town I was in. I really had it nice there. I was in a house and had a swell stove to cook on. The last night we were there they got the electric power plant going again and believe it or not we had a picture show in one of the hotels there. It sounds like I was on a vacation, but actually the front lines weren't but about 2 miles away.

Right now I am living in a big bomb proof shelter. We receive quite a bit of artillery fire from the Jerry's here (one just came over right now as I wrote that, but it exploded a little ways from here). This shelter comes in plenty nice, but it gets kind of rough when they blow up our telephone wires, and then we have to go out and fix them. Boy, I am sure getting jumpy lately. I am ready to hit the ground at the slightest sound. Outside of sweating out those artillery shells we have a pretty good deal here. There are 9 of us and an officer here, so we are drawing plenty of rations and we have elected one guy as cook and that's all he has to do.

We also have a short wave radio here to use for communications when our lines get blown out, but when all our lines are in we use the radio and get all the first class news reports and plenty of good music. "Wow", I was writing this letter outside, but a shell came over and exploded just a little ways from here and now I am way inside this shelter writing by candle light. I just found out that the last shell landed right on top of a house I was sleeping in the other night. That was one of the worst nights I have spent in a long time.

The house was so crowded I and two other fellows slept under the bed. The Jerry's were laying mortars and artillery shells all night. They would let up just long enough to where you would just be on the brink of dozing off to sleep and "wham" one would land right outside and plaster would drop off the ceiling and the shrapnel would beat against the side of the house. I was just waiting for one to land right on top of the roof, but none ever did. The way they were coming down around us I thought for sure they were going to climb right in bed with us.

Now I am operating the switchboard, say I am really taking you on a tour, aren't I? I have taken lots of pictures so I will be able to show you all the places I write about. We even had a cat for a mascot, but he went AWOL (absent without leave) when a few shells came over. Well, that's the daily routine around here. Get up in the morning, eat breakfast, then fix the lines that go out, come back and take it easy listening to the radio and writing letters until another line goes out. Oh yes, we also take turns operating the switchboard. It gets pretty cold inside of this shelter. It's built right under a mountain. I don't like to wander too far away from it though unless it is absolutely necessary because you never know when a shell is going to come over. I would rather be on this side though, because our own artillery throws back many times more to Jerry than they through into us.

The letter ends here with no signature

Letter to Mom from Harland "Bud" Curtis

Southern France, Col de Braus, Sunday, September 24, 1944

"Viva La France"

Dear Mom,

Just a short letter today to let you know I am okay and I am still hanging my hat in this air raid shelter. Haven't done much today. Jerry (Germans) threw over a few shells this morning, but none of them hit our telephone lines this time, so I have been taking it easy listening to the radio. Looks like we aren't going to get any mail today either. I hope we start getting mail again soon. It has been a long time since I have heard from anyone. Not yet have I got a letter that you know I am in France. Yesterday though, I did get that super glamorous picture of Jill. It really is a swell picture of her. Just exactly what I have wanted for a long long time. It sure brightens this ole tunnel up. Try and get another one just like it and put it in my room for me so I will have a nice one when I come home. This one I have here has a lot of travels ahead of it and anything can happen to it. I wish I had a small one to put in my wallet also. I have all kinds of pictures of her, but I never get tired of them and I want more, so any time you want to make me happy, just enclose another picture every time you write. Does Jill get much chance to come out to the house anymore or does school keep her pretty busy now? Take good care of her for me. How is Lorrain and Harley? Tell Pop hello and everyone else. I guess Bert is well out to sea by now. Gee, I wish I would get some mail. Oh I just about forgot to mention I got another one of those small packages. They are sure tasty. Thanks a million. Keep 'em coming and send some stamps next time too. Okay.

That's about all for now. I'll write again soon.

Lots of love, Bud

Tunnel at Col des Braus, France, 1944. Bud with mascot "Adolph the cat". Bud said when the German shells came in, Adolph could hear them coming before we could. When Adolph started running it was a signal for us to run into the tunnel for safety. In Bud's letter November 4, 1944 he explained the Artillery guys "catnapped" Adolph and he never returned. Adolph got his name from his long mustache like whiskers. In June 2004, Bud returned to the tunnel where he had lived sixty years earlier.

Letter to Mom from Harland "Bud" Curtis

France in the tunnel of Col de Braus
4:45 a.m., Saturday September 30, 1944

Dear Mom,

  I am sitting here at the switchboard inside our tunnel writing this by candle light. We take shifts throughout the night on this switchboard. Someone has to be here all the time. In an hour I will wake up Joe and he will take the last shift until 7:30 a.m.; then we all get up, get breakfast and that is the beginning of the day. We have all our telephone lines in good now and the artillery shells the Jerry's (Germans) throw in at us haven't been blowing them (phone lines) out as much as they were when we just came here, so there is really not an awful lot for us to do except listen to the radio and read and take it easy. They let a few guys at a time have a 2 day pass to go into Nice. My turn isn't too far away. I had to go to a dentist the other day and the hospital is there in Nice, so I got to see a little of the place and from what I saw, and from what the fellows who have been there on pass say it seems to be a swell place, even better than Rome. I will take some pictures of Nice when I get to go there. I have gotten some films developed there of the pictures I took since I have been here and have mailed them to Jill to put in a scrapbook. I hope to have someday after the fellows get through using the negatives to get more pictures made from them I will send them on home and you can get some prints made there in Long Beach. I have about 10 or 11 rolls of film now that I bought in that last little village I was in. The one I told you I was living in a nice warm house and how swell everything was. I liked that place – wish I could of stayed there for the duration. I have some pictures of the place, so when I get those developed, I will send them on home or to Jill. I wish I had some way of sending you some film as there doesn't seem to be any shortage of it over here. I wish you would send a picture in every letter.

  We still haven't got any mail – better than two weeks now. There is something messed up farther back the line, but they had better get it straightened out pretty darn soon or you will be hearing of another war between the 517th and who is ever responsible for us not getting our mail. I have plenty of time here to write letters, but it's awfully hard to do, when I'm not really doing anything now to write about; and I don't even have any of your letters to answer questions for. I guess you know what I mean because it goes sometime every so often that you don't hear from me. I have been using most of my spare time to study French from a couple of books I have. I understand more than half of what I read when it is just written down, but I haven't had very much practice in speaking and pronouncing the words. Today I had to ask a lady twice for potatoes before she knew what I wanted. Very discouraging after all my studies. I guess I kind of murder the language, but I get my point across anyway. See if you can find an English – French – French – English dictionary there in town and send it to me in your next package, will you? I learn a lot of words by just pointing to something and ask them what it is, but some things you can't point to. That is I did in that last village I was in Peira Cava – but I don't get much chance to talk to anyone where I am now.

  I only have 3 stamps left, so I hope to get a letter from you soon with some stamps or I will have to use V-mail *(V-mail was a preprinted military style letter, more like a post card that could be sent for free)* until I can get some more. I am enclosing a receipt for $95.00 I sent home, so be looking for it and that $45.00 allotment I have made out to you. It will be pay day *(only once a month)* again in a couple of days and we get $30.00 back pay on that *combat badge (Combat Infantry Badge awarded to infantrymen once in combat for six month and each soldier received $5.00 per month for each month in combat)* we won fighting in Italy. It pays $10.00 every month.

  That's about all for now – I hope everyone is well and give my love to all – take care of Jill for me and have her come out to the house as much as she can, if it is possible now with school starting again.

Irene, says she will stop by and say hello when she comes to California. Maybe you wont like her, but she has been awfully nice to me. Just like a big sister, so treat her nice and tell me when she stops by.

Love again

Bud

Bud just out side the tunnel in Col de Braus, France, 1944. Bud washing up with his G.I. helmet on a small field stove at the tunnel. Picture of small field stove taken at the Liberation Museum, Le Muy France, June 10, 2004

Letter to Mom from Harland "Bud" Curtis

Col de Braus Southern France
October 6, 1944

Dear Mom,

At last I finally got some mail from you and Jill and another from you. First mail I have had since about 3 weeks. Gee, it was sure swell that Bert was home to get the baby christen. So those silk stockings finally made it to Jill. I was beginning to wonder if they were ever going to get there.

I have our pet cat Adolph sitting on my lap and he is getting just a little too frisky for his own good. If he makes another swing at this pen I am going to throw him out side in the rain. Oh yes, it has been raining all day. Then last night a dog walked in here to our tunnel, so we have a regular Noah's Ark here now. Adolph doesn't think much of the dog though. Boy last night as we were starting supper the Germans laid down a terrific barrage of artillery and caused us about 19 different kinds of hell. They blew out everyone of our telephone lines to all our units. They had the shells timed so that when they got right over us they would explode about 10 feet off the ground, and what I mean is that it is plenty dangerous to be any where around. Just the concussion about knocked us out. They were landing so close to the front of this tunnel.

Gee, communications got so bad and it had turned completely dark we couldn't wait any longer. We were hoping they would ease up a little bit. Everything depends on these telephone lines. They run to all of our units. Our outposts. Our artillery. Our medics. Our headquarters, and everyone of those lines had been blown to a million pieces. What a rough job it was going out there in the middle of all those shells, and it was so dark you couldn't see a foot in front of you. We just had to run the wire through our hand as we walked until we came to a break then splice it back together. When I walked out of this air raid shelter *last night I honestly never expected to walk back again alive, but once more my luck held out again. One of these days my luck will catch up on me. 52 **days** on the front lines and not a scratch so far. Some days are better than others. Sometimes, I am as safe as if I were right there at home and other times like last night I don't think ever the Lords of London would have bet 2 cents on how long I was going to stay alive. Today has been very quiet outside because it is raining. Not one shell has come over all day. That is the main trouble with that artillery. You never know when they are going to open up on you. I try not to wander too far from this tunnel, but when a line goes out, it has to be fixed and that's my job and a darn important one too. It's those darn mortar shells that are dangerous because you can't hear them coming until they are right on top of you. At least that big artillery gives a guy a little warning as you can hear the shell scream as they come and it gives you time to hit the ground. This life I am leading either drives a guy nuts or else he gets use to it. I'm just in the middle waiting to see which way I am going to go. Sometimes I walk out of here and don't give a darn if they put a shell in my back pocket and other times if a fly buzzes a little too loud I am ready to hit the ground.*

Lots of rumors going around now. Some are of us going back to the States for a while and then to the south pacific. Others are just of us going straight there and others are of us staying right here. I guess no one really knows. Boy, if they send us to the South Pacific I'll give up all hope of ever seeing home again.

In about another 2 weeks or less, if we are still here, I will get a two day pass to visit Nice. I was in Nice for a little while one day when I had to go to the dentist and get my teeth fixed. It is really a swell place. I like it even better than Rome and I have been studying this language for the last two weeks and can actually talk to these people and it sure makes me feel good to know all my studies haven't gone to waste. I have a dictionary now, so I can look up words I don't know or else I just ask them how they say it and then try to remember. I met a girl there who was born in Columbus, Ohio and lived there until she was 10, then they came back here to France and she is 18 now.

*She remembers about as much English as I have learned of French, so between the two languages we got along fine. I got some more films (**pictures**) developed and have sent them to Jill, so there are quite a few pictures on the way now. What size film do you use? Maybe I can get you some over here. I have about 11 rolls for my camera so don't worry about sending me any. Just send me pictures you take there. If you want to send me something for Christmas, I could sure use a pen. I lost the one I had and have to borrow one every time. I will try and get Jill something here, but I will have to pay twice as much or more for something over here. A 25 cent comb costs about $1.50 in our money. I'll be sending some more money home so if I don't get her anything here, buy her something real nice and get something for you and Dad, and Bert and Lorrain. It's going to be hard for me to send things so I would rather you do that. I don't care if you spend every cent I have sent home. Money doesn't mean anything to me so get something real nice. I would sure love to give Jill an engagement ring for Christmas, but if I did she would have to pick it out herself and I would want it to be the best. I don't know exactly how to go about doing something like that by mail so if you have any suggestions please tell me. I don't know if it would be the right thing to write and ask her or not, so maybe you there at home can figure it out and then let me know what to do. I haven't even mentioned it to her and maybe she or her mother still wouldn't like the idea because she is still going to school, and for all I know maybe she wouldn't want to anyway. She never mentions anything like that in her letters. Oh well, it was a thought. Maybe I am on the wrong track again. Let me know what you think?*

Gee, it gets colder every day. I expect every day to wake up and find snow on these mountains. It is about 6:00 o'clock and time to eat. We have elected one guy to do all the cooking and that is all he has to do (H. G. "Sug" Lawrence). There are about 10 of us here.

Eight now. 2 are in the hospital, so we are having a little more to eat. I don't have the best of food, but I'm not going hungry.

(Time Flys)

It is quite a bit later now and I have finished eating and I am now back in our tunnel where I sleep. I have four Kerosene lamps burning and it is as light as day. When we first moved in here it was damp and cold. Put your hand on the wall and it would be dripping wet, but now it is dry as a bone and a heck of a lot warmer in there than outside. It was hailing the last time I looked out.

We found an abandoned car on the highway and drug it up here with our jeep. The axel is broken on it so we have taken all the seats out and have them in here. Right now I am sitting on one of the seats. Sure nice and comfortable. We had the car sitting right out front and it was really torn up from the shrapnel from the shells Jerry (the Germans) dropped on us. You ought to see the size of the dud that landed out front. If that thing had gone off I think it would have knocked us all out from the concussion. It is sure a wicked looking thing. Wish they were all duds like that one. It's been awful quiet today, they haven't fired a shot or neither have we.

Well that's about all for now. Write often

Lots of love, Bud

Letter to Mom from Harland "Bud" Curtis

Southern France, later to be Col de Braus
October 11, 1944

Dear Mom,

Got some more mail tonight and it is sure swell to hear from everyone again. We didn't get any mail for about 3 weeks and now it is beginning to drift in.

I got that 2 day pass to visit Nice. I was telling you about, and what I mean, is it sure is swell to get away from here to a big city. I have never felt so much at home in any place as right there in Nice. It is on the coast just like Long Beach and down town there are big modern department stores and all in all Nice is nice (Nice is pronounced as "Niece"). *I stayed at a very nice hotel and was it ever nice to sink down in a big soft mattress, clean white sheets and big pillows. I no sooner lay down than I would be sound to sleep. The 517th has a nice hotel where we go to eat and they sure fed us good. It was a regular vacation. I wrote Jill a long letter tonight and told her all about it.*

"Adolph" our cat, is getting a little too frisky for his own good. Notice the holes at the top of the page. He took a swing at my pen and missed. There is a million and one places for the little devil to go in this tunnel but he insists on occupying the lap of whom ever is operating the switchboard. He is a cute cat, but he seems to think he owns this place. You should see him make for this shelter when Jerry throws in the artillery.

Got a letter from Bert today. Wish I could send him something for his birthday Friday (Bud's brother Bert was born October 13). *He said he had just docked at_____* (and there it had been cut out, censored). *There is an ex-paratrooper on ship with him that got hurt at Fort Benning. I don't know him, but I heard about that accident so I know of him. Chances are he might have been right here instead of where Bert is.*

We got our duffle bags now and I have lots of warm clean clothes again which makes me very happy. I have some pictures being developed that I took in Nice and some duplicates of some others, so I will be mailing them out your way soon. I hope I get that package with the cake, pen and pencil. I can use them all to a great advantage. I am waiting for your opinion of what to have you get Jill for Christmas from me so write soon. We should get paid again soon now and if I get to go to Nice again which I think I will, I will get her something, maybe some nice perfume.

I met a very nice girl in Nice (Jackelyn) and she was born in Columbus, Ohio, and then they came back here when she was 10. She and her brother remember a little English so with what French I have learned and the English she knows I get along fine. Jackie's mother lived in England and America for a long time, so she speaks perfect English. It is sure swell to have people treat me so nice as they do. Her mother even said if I will let them know the next time I can come to Nice, that she will make an apple pie, and that's something, because food is kind of scarce over here. That Jackie is a wonderful cook. She made me a chocolate pudding and it sure tasted wonderful. Her brother is an intelligent kid. He is 18 and likes radio and electricity and is swell at mechanical drawing. I wish I had more time to spend there as I would like to help him build a radio just for the fun of it.

They are a darn swell family and it is sure nice to have a place to go like that and be treated so good. I seem to meet people like that where ever I go. I really like these French people and if I was ever going to live in another country it would be right here in France. It wouldn't take long for me to learn this language either. I can understand most of what I read, but I still need practice on listening and speaking. I got along fine in Nice and it makes me feel like my studies haven't been wasted.

It is time for me to wake up the next guy for this switchboard so I'll close shop. "Adolph" is going to get plenty mad at me. He is sound asleep now. He is even snoring.

In one of your packages how about sending some tooth paste if you already haven't. Thanks a million. Tell everyone hello. Lots of love,   Bud

Letter to Mom from Harland "Bud" Curtis

Southern France, later to be determined Col de Braus
Friday, October 13, 1944, Bud's brother Bert's Birthday

Dear Mom,

Wrote Bert a long letter tonight. Sure hope he gets it, but I also included your address so it will be sent home if he doesn't get it. I sure wish I could send him something, but at least he will know I hadn't forgotten about him if he gets that letter. It has been a long time since we were together on a birthday or Christmas hasn't it. This war has sure changed Bert and my paths in opposite directions alright. It has been so long since I was home for any length of time it is hard to believe he is now married and has a cute little kid. I still think of him as when we were both just kids. He is the best brother a guy ever had and I don't put anyone above him or you and Dad. Guess I have been pretty lucky all of my life.

It has been raining all day and our tunnel is about 4 inches deep with water now. Haven't been doing much. Days come and go and run into each other and I don't hardly remember what I do from one to the other. Doesn't matter much anyway I guess.

Gee, how I wish that I could be back again to the days when Bert and I were just kids. Had lots of fun alright we did. Can't make me forget things like that no matter how far from home I get. Sometimes those memories seem just like a nice dream I had a long time ago and that they really never did happen. That's what this way of living will do to a guy though. Seems like I've been living in caves and dodging shells all my life sometimes. It is only when I get letters or I am writing them that I realize there is really another part of this world where people are still living like human beings and don't really know there is a war going on.

Haven't much to say, just sitting here at the switchboard with nothing to do and thought I would drop a line and say I am okay. Getting kind of sleepy, been sitting here for 7 hours now. Just writing letters and thinking, and watching the water creep up higher on the wall. Everything is quiet as a tomb except for the sounds of dripping water and some shells exploding outside. Enough to drive a guy nuts. I guess I will turn the radio on. Got a program of about 100% corn from London, (corn meant it was "corny or silly") *but at least it is something so I will close this letter and listen to it for a while. Haven't any more to write about anyway.*

Love,

Bud

P.S. Got a letter from Irene today. She will be out there pretty soon. She is a swell girl. She has sure treated me nice.

Letter to Mom from Harland "Bud" Curtis

Southern France, later to be determined Col de Braus
Tuesday, October 17, 1944

Dear Mom,

It has been a pretty nice day. Sun has been out nice and bright and has been really nice. It will be raining again though, as winter is coming on. Guess it won't be too much longer until I come out some morning and find snow all over these mountains.

Haven't had any mail for about 3 days, but I hear they have a lot of it for us whenever they can get it here. I guess Irene (Older girl Bud met in New York City while on pass from Camp Mackall. He thought of her as an older sister) *will be out in California by about the time this letter reaches you. She is a darn nice girl and has sure been swell to me. She only has one bad habit and that is she smokes too much, but outside of that I think she has done pretty darn good at taking care of herself all these years all by herself. Her mother and father were killed in an accident soon after they came to America from France and she has been more or less on her own ever since she was a kid. It sure makes me feel good to know I have friends when I am in another city and a long way from home; so whenever she is in Long Beach I will appreciate it very much if you will be as nice to her as she has been to me when I was in New York. If it isn't too much trouble maybe sometime you can invite her to stay for a couple of days. She could stay in my room, but of course that is up to you* (Irene did come to visit Bud's folks and did stay for a couple of days. Shortly after the visit to California, Irene returned to New York City. The last time Bud ever saw her was on his pass in 1944 from Camp MacKall, just before shipping out for overseas).

Haven't been doing much the last few days. Jerry (Germans) *has been rather quiet. They didn't shell us at all yesterday, but threw in a few today.*

I will have a one day pass to visit Nice again pretty soon and I will get some more pictures. I have finally got some pictures I took in Italy being developed and one of the fellows on pass now will bring them back tomorrow. I am enclosing a couple of reprints from the pictures I sent Jill. I guess you have seen these by now, but here is some more anyway.

That's about all for now, will write again soon.

Lots of love, Bud

In the next letter (October 22, 1944) Bud told his Dad that he got a pass to Nice. What he didn't tell his Dad was when he got back to Col de Braus he faced a court martial. Bud explained, "It was customary when the troopers received a day pass to Nice they stayed overnight, and came back to camp the next morning. Nothing was ever said to the troopers about staying overnight, and it was an acceptable practice. On this particular day the two troopers allowed to go to town for some reason did not or could not go. It was announced in late morning that two troopers could go to town. At first I did not want to go because half the day was gone, but I knew I could stay overnight. I got excited about going to town, and so I asked Sug Lawrence (another trooper in the communication section) to go with me.

He agreed, and so we got on the truck and left for Nice (40 miles away). We had a great time in town and a great evening. We stayed in one of the many hotel rooms, and came back to Col de Braus late the next morning. What we did not know or we forgot was it was "Payday." Every man was to receive his pay in French Francs. Our names were called, but we were not there. We were technically supposed to be back in camp that night, but we stayed overnight just like every other trooper had done. This apparently was against regulations, but everyone did it so why couldn't we? The only reason was "Payday" and all troopers were required to be there to receive their pay. We were not there, and so we were "AWOL." Absence without leave. We were now in serious trouble. Sug and I would face court martial.

    Our Company Commander, Captain Ehly recommended us for court martial in October, but we were in a combat zone, and had to wait until we got back to garrison. This meant Sug and I had to wait until November when we finally arrived at La Colle-sur-Loup, six miles west of Nice. Some time in November 1944, Sug and I were brought before a court martial board. The court martial board did not go easy on us, and I believe wanted to make an example out of us so the rest of the men would not stay overnight in Nice. Sug and I were confined to the camp, and were not allowed to go to town for thirty days. We also forfeited one half our pay for six months, and were given thirty days of extra duty. It seemed like a life sentence. We cleaned every dirty pot and pan along with any other dirty detail the sergeant could think off. We were upset, and felt we were not dealt with justly, but that is the Army (In February 2006, Bud and Sug Lawrence spoke on the telephone after 62 years. A major portion of their conversation was about that court martial. To this day they are still upset, and feel they were dealt with unjustly). On the weekends all of the men would go to town. There was no one there to watch me so I took off for some of the smaller towns in the local area. There the French people welcomed me with open arms, and I believe I had a better time there than the rest of the guys did down in Nice. I was never caught going off for the weekend jaunts."

Letter to Dad from Harland "Bud" Curtis
                                Southern France, later to be determined Col de Braus
                                                            October 22, 1944

Hi Dad,

    I'm still here in this air raid shelter and I am about ready to hang up a sign, "Home Sweet Home." I have been here for so long it is beginning to get a bit cold in these mountains. I expect anyone of these days to come out and find snow all over them.

    The Krauts were playing a little rough yesterday. I was beginning to believe that they couldn't get these shells much closer than they have been, but they fooled us yesterday by putting some of those big ones that sound like a boxcar coming at you right practically on top of us. I thought for awhile it was going to get all of us, but all it did was about jar our teeth out. We got a cook shack about 100 yards from our tunnel and was just getting supper ready, but when the concussion from one shell blew out a new window we just put in and put shrapnel almost through a foot of the concrete walls, we decided it was about time to get out of there and into this tunnel. We just did make it in time. They couldn't get us in here with blockbusters. They blew out most of our telephone lines so we were kept pretty busy for a while getting them back in again.

    I got a pass to go into Nice (The word Nice was scribbled through, perhaps by a censor who read the letter before sending it on to the States) *the other day. It is a swell city build on the coast the same way as Long Beach. Everything about the city is modern except there aren't many cars over here and the ones they do have over here have to be fixed up with a boiler like contraption that they burn charcoal in to take the place of gasoline. Most everyone rides a bicycle and they are about as thick as cars are in America. I met a girl in Nice that lived in Columbus, Ohio for a few years when she was young. She still remembers a little English and I have learned enough French to get by; so between the two languages we got along fine. Her mother speaks perfect English and while I was there they made an apple pie for me and did it ever taste good. It is swell to have people treat you so nice. I told her when they are allowed to write letters again to America for her to write me there in Long Beach and if I wasn't there that you would send the letter on to me.*

I would sure like to have gone on that fishing trip with you and Hy (Hyrum Douglass Bud's uncle). Boy those fish must have been standing on each other from what you said. Maybe one of these days, you, Bert, and I can get together and take off on a good ole fishing trip again. I would sure like that alright. I'll bet you can still out hike the both of us. That holding down two jobs must be keeping you pretty busy. I got paid the other day and as soon as I can I will send the money home. I think I can send a $100 or more this time and watch for that $45.00 allotment each month that I had made out to be sent home. How much have I got there now anyway? It really doesn't matter. I want you to use it for what ever you want, and get Jill something nice from me for Christmas, and also for you too. I would have to pay 3 or more times the price for anything over here and you can get better things right there at home. The only way you can get any good out of your money here is by sending it home.

Wow, am I glad I was in here just now. Jerry (Germans) just landed a couple about 20 yards from the door. We had some clothes hanging on a line out there. A guy just brought them in and everything is torn to shreds. A couple of our guys were down there at our cook shack. It is a wonder they both didn't get killed. We have all been pretty darn lucky alright. Our artillery is giving them hell now and they haven't quieted down a bit. I hope they don't put anymore in here until after chow because we got a chicken in our rations this time and I don't want anything to keep me from eating my first chicken for many a month.

How is Harley doing now and what do you hear from Bert? Sure wish we could all be home for Christmas, but maybe by next one. Sure hope so anyway.

That's about all for now. Tell everyone hello and I will write again soon.

Lots of Love, Bud

P.S. Those paratroops you said you saw in the News Reel, that jumped on Southern France wasn't us. We jumped <u>early</u> that same day when it was still dark and foggy. When those guys you saw jumped in that News Reel came in, we had already had that whole territory in our hands and I was on a high hill at that time and watched them jump and had a perfect view of all the gliders coming in. Our Lieutenant took some swell pictures of those guys jumping and of the gliders. When he gets out of the hospital I'll get some copies made. Those guys were really lucky. We already had taken everything around where they jumped and then they were only with us for 3 days and were pulled back and have been in a rest area ever since. I am beginning to doubt if we will ever be relieved.

Letter to Mom from Harland "Bud" Curtis

Col de Braus Southern France
Tuesday, October 24, 1944

Dear Mom,

Some of the old mail is beginning to drift in now. Got one from you with a picture of Bert, Lorrain, and Harley. Two from Jill, one from Garth, Roger and Frank Hill and one from Jerry. She sure is nice about writing and I really do like her letters.. Tell her I said that. Got that Caronet magazine and the stationary you sent. I have plenty of stationary now and envelops. Just now and then send a airmail stamp. Send all the magazines you want as they are well appreciated by me and my buddies. How's chances of an Esquire now and then. I'm sure the pin up babes would be very much appreciated to brighten up the wall of this tunnel. Don't send any more canned things as I have plenty of that now. If you can find any powdered milk anywhere, send that. My teeth are about ready to fall out from the lack of milk. Send all the cookies and candy you want. We don't get things like that either.ABig, a big shell just landed outside the door and my ears are sure ringing. Had a couple of close calls today. They are getting a bit rough this past week. That last shell just knocked out one of our telephone lines. It isn't going to get fixed now though. It is 3:00 o'clock in the morning and it is raining cats and dogs.

Don't worry about that package I said I sent from Camp Mackall. The guy who was suppose to have sent it for me is dead now anyway. He was electrocuted cutting a high tension power line on D-day (During the invasion of Southern France, Bud recalled this incident. In April 2004, Bud remembered that this soldier was ordered to climb up the telephone pole and cut the power line. The soldier objected and was afraid of being electrocuted. The sergeant or officer in charged ordered this soldier up the pole to cut the wire. He obeyed the order and was electrocuted. It cost him his life).

It is time for me to wake the next guy up to watch this switch board and I am getting sleepy, so I'll close for now and write again soon. Thanks for all the packages you have sent. Don't worry about the cold, I have a lot of warm cloths now, but can use whatever you sent anyway. In one of your packages include a bottle of Vicks nose drops. I might need it sometime.

Lots of Love, Bud

Letter to Mom from Harland "Bud" Curtis

Col de Braus Southern France
October 28, 1944

Dear Mom,

I guess I have received most all of the things you have sent in the big envelopes. I have plenty of stationary for now, but if you want to I would like very much if you would keep sending magazines, that way as I sure like them. That Jeevers book you sent is sure interesting and also the Caronett and Digest. Books and magazines are well appreciated by everybody and when any of the guys get them they circulate around until they are completely worn out.

I guess I told you I wrote to Bert. I sure hope he gets my letter, but I put your address on it also, so if it misses him somewhere it might end up at the house. It has been wonderful weather the last couple of days. I have lots to write about, but can't for a couple of days.

That is the way it is over here. Never can say what is going on or what is go to happen. Just have to wait until it is past news.

I have sent a few more pictures to Jill. Maybe I can get some film for you here. I would sure like some more pictures.

Had pork chops for chow tonight and they were really delicious. Have plenty to eat now so if you send anything make it cookies, candy, toothpaste. I have plenty of soap. I guess you will have most of those things in the packages you have already sent.

About all I have to say for now, but will write again soon.

Love, Bud

Letter to Mom from Harland "Bud" Curtis

Col de Braus Southern France
October 30, 1944

Dear Mom,

Got your October 19th letter today, but as yet haven't received the one you speak of which you explained my asking you what I wanted you to get Jill for Christmas. I am so far away and have been gone so long that it hardly seems like I ever was any place but right here anyway. Not much matters to me one way or the other anymore. My hopes of ever getting back to the States again are very low in my estimation, and if I do, I will just consider myself lucky. You people keep going on as you always have, but over here things are changing and a guy just naturally changes with them.

The Krauts moved back from this position they have been holding and soon I'll be moving up again. I went down to the position I spoke of to check a telephone line today. It was a war torn place alright and the people were very glad to see the Americans. It was kind of rough for the people there when the Krauts held the place. The wouldn't let them be evacuated and they had hardly nothing to eat and all the time our artillery was shelling the place to bits. All day people are walking along the road carrying whatever they can with them on their way to Nice and plenty glad to get out of that place. The people in the States just don't realize how lucky they are. They think the war here is over. Just wait and see. It is far from finished yet. The Germans had to leave in rather a hurry and there was lots of equipment and a few dead Krauts still left along the road. I am far from being a souvenir hunter, but with so much junk laying around it is hard not to pick up something. I am enclosing a couple of pictures and a German insignia I picked up.

Before I received this letter from Jill saying not to send any more things I had already sent a German helmet and some parachute silk I have been packing around since D-Day (Deployment Day). I doubt very seriously if she will ever get it as with what things I had to use for a package it was rather a rough job. If she does get them, well she can throw them away. It doesn't matter.

It finally got around to snowing last night. Quite pretty, but awfully cold. I guess we are in for a rough winter. I made some ice cream out of snow last night and it turned out pretty good.

I didn't realize I had sent so much money home. I will have a $130.00 money order on the way soon and also that $45.00 allotment should be there about now. This is the end of another month today, so I guess there is another of those $45.00 allotments on the way, and when they pay me, I will send what ever it is on home. Money is of little use to me, so you use it for getting everyone something nice for Christmas. Spend it all. I don't want any of it.

Hope I have made all my points clear and not offended anyone.

Love as always,   Bud

Letter to Folks from Harland "Bud" Curtis

Sospel France
Saturday November 4, 1944

Dear Folks,

Sorry not to have written for so long, but with moving up to this town that the Germans pulled out of, and then having to put in new telephone lines again I have been going around half fed and half asleep the last few days, but now things are running smooth again and when I stop to look around me, I notice that this is a pretty nice set up. I and three other guys have a room in a building we took over and it is pretty nice. We have our switch board in the closet and a stove in the corner and two bunk beds, mattress and everything.

Probably last week some German was sleeping on this same bed, but I am glad he moved out because it is sure heaven to that cement floor I was sleeping on for the last month in the tunnel. This place was kind of a mess when we got here, but is all nice and clean now. I am sure glad I wasn't here on the receiving end of our big artillery guns, as they didn't miss anything. So far the only place the Germans have shelled since we have been here is way out of town. They are probably doing it by guess work, anyway it has been pretty nice not to have them land in our front yard like they were up at the tunnel (in Col de Braus). Your life wasn't worth much at all up there.

I got a package from you the other day and one from Loa. This is some of the stationary she sent me and she sent enough to last me the next 6 months, so don't you go to any trouble of sending more as I won't know what to do with it all. I will write and thank her tomorrow. In your package was the marshmallows and cookies. I have plenty of can stuff now. Most the guys did just like I did when we first got here to France and were eating K and C rations. They wrote home for a lot of canned stuff and now we have plenty of it. They are still getting more in the mail. From now on just send cookies, marshmallows, candy, and fruit cocktail and things like that because these are the things a guy is glad to get any time no matter how much there is to eat.

We even get fresh butter, bread, and chicken, turkey, pork chops, steak and others quite often now. We have to cook all our own meals, but that's the best way to do it. We have a guy here that use to be a cook and he takes care of all that and is he ever good. I would rather have it this way any day than sweating out a long chow line. A guy could live forever on these 10 in 1 rations we get, and you can eat whenever you want. I really am satisfied with the whole set up, and at times it doesn't seem like your even in the Army. I don't know what it seems like because it is just something different to just take over a town and the whole outfit moves into people's houses and have big beds to sleep in. A fireplace, a kitchen, and cooking their own meals. Well it just ain't Army, but it is lots of fun in a way and you can almost forget at times there is a war going on.

I sure wouldn't want to be living in any of these countries after the war. These people will be getting killed years after the war is ended by mines both we and the Germans have planted everywhere and are not dug back up.

Just today some of our guys were killed by a time bomb, and the other day our ex-First Sergeant, who was made a Lieutenant just a while back and everything was going along fine for him. He and another fellow walked onto some mines and it blew a leg off each of them. Those are the awful things about all this. Two good guys like that being crippled the rest of their life. Of course this is war and all that, but I'll never get use to hearing about things like that and my buddies getting killed. It's all so darn useless and silly, but those things will just have to be until the Germans decide they've had enough. Even as much as I hate all this and would like to be home, I would rather go on fighting the next 10 years if it meant it would end all this forever. I never want to have any kids of mine going through this same thing 20 years or so from now, and I guess all the other guys feel the same way about it although just one man's life in my opinion is an awful high price to pay for that. I'll sure be glad when its all over and I can forget all of this.

Those darn artillery guys that were living next door to us at the tunnel "Cat Knapped" Adolph our cat when they moved out, but I'll get him back if I ever see him again.

We I guess it is time for me to turn off the midnight oil I am burning here, ah reckon, so good night for now.

P.S. I will write the Borax Co. tomorrow and see if I can't get that check made out to Pop and have it sent to him. I hope Jill is feeling better. I am glad you heard from Bert and hope he gets home soon. Has Irene come by yet? Thanks again for the package, and now I really will stop with all my love to all.

Bud

Oh, Oh, another P.S. I got more of the magazines you sent and was really glad to get them. That is really a good idea and a fast way to get them here, so keep on sending them. The guys and me too would like some Esquire Magazines so we can have some pin up pictures. If it isn't too much trouble send al the things like Coliers, the Post, those books - American – Popular Mechanics (I received the ones you sent, thanks). They don't particularly have to be the newest ones out because most all of them for the last 6 months back are new to us here.

Thanks again and this time I really, really will say good night.

Love again,
Bud

Letter to Mom from Harland "Bud" Curtis

Sospel, Southern France
Wednesday, November 8, 1944

Dear Mom,

Have received two packages from you in the last 2 days. Right now I am writing with this pen you sent me, and as you can see it writes swell. Thanks. I sure have been wanting a new pen and the ones you buy over here aren't much good. The packages got here in fine condition and as for the cookies, candy, etc, they tasted like you had just put them in and handed them to me everything is so fresh. I was sure glad to get this honey, it has been an awful long time since I've had any, and does it ever taste good on the salt crackers or on toast we make on the stove.

Things have been pretty nice here in this town we are in. We haven't been shelled much to speak of, and it is sure swell to have this nice room. I'm sure if we had stayed much longer in that tunnel we would all have had rheumatism or something. My teeth have been bothering me quite a bit lately, and I went to a dentist, and he said all that was wrong was the climate was affecting them, and that it is nothing to worry about.

Got a letter from Afton (Bud's cousin) today, and she says she is very happy with being married and that she is going back to be with Bob (Bench), soon now. That should work out fine I guess, because she says if everything goes right that he will be there for the next 2 ½ years with that V-12 program (The V-12 was for men who were in medical school and paid all expenses. Bob Bench became a doctor and later served in the Army). That's pretty nice for those guys. Much grass will grow in 2 ½ years.

I guess if I had it all over to do again I would go into the paratroops, but I still can't help but envy those guys that are still in the States, and probably will never know what this war is all about. I have had many experiences and have learned a lot since I joined this outfit. Most of what I've learned has been the hard way, but you don't ever forget anything when you learn it the hard way. I can truthfully say I don't regret coming into this outfit, but I will sure be glad to get out of it. I don't ever as long as I live ever want to make another combat jump or any jump at all, but of course if it is to be done, I would never refuse to do it. I'll never refuse to do anything. Those combat jumps really put the years onto a guy's age. It is hard to explain what kind of a feeling it is to step out into space and not know what is down there waiting for you. I'll never forget that day as long as I live, but it can't be explained in words. I went to a picture show they had here tonight. It was Red Skelton, "Whistling in Brooklyn." First show I have seen for quite a while.

I am glad to hear Jill is feeling better now and I hope she stays that way. I want you to get her something extra special nice from me for Christmas. I will soon have money orders on the way for $210.00. I hope the $45.00 allotment for the last two months gets there okay. Jill says she just loves new suits and clothes. I guess most every girl does, so why don't you just take her downtown with you and let her pick out for herself the best one there is in town.

Write soon and tell me what you think is best. Thanks again for the swell package and this pen and pencil. Send some more cookies and candy if you want, I sure like to get them. Lots of love Bud

Once the American forces gained control of Col de Braus the 517th marched to Sospel. Bud remarked, "It was about twenty miles to Sospel. As we march down the mountain about ten miles we stopped for the night and stayed in some houses along the road. As I entered the house I saw two German packs. I picked these two packs up and the next day carried them all the way to Sospel and then to La Colle-sur-Loup. When we got to La Colle-sur-Loup, I was given my duffle bag in it was my other uniforms and boots. I changed into clean clothes, and then I put the German packs in my duffle bag. I then returned my duffle bag to our supply sergeant who sent it off for storage. I did not see my duffle back until after our campaigns in Germany were over. On my way back to the states I kept the German packs in my duffle bag and brought them back home."

In Sospel conditions for the civilian population deteriorated rapidly. The Germans cut their food rations to one and three-quarter ounces of bread a day. The Germans took thirty French men and forced them to work in dismantling installations and evacuating supplies. On October 20, 1944, the German commander announced that the Sospel population was to be evacuated to northern Italy. The French refused to go, and on October 27, 1944, the Germans blew up all bridges including one that was built in the 11th century and fled the town up into Italy. The Americans arrived in Sospel the next day.

On November 4, 1944, a house sheltering seventeen 517th troopers from the Second Battalion blew up killing five men and wounding five more. The building was either booby-trapped or the Germans had left a delayed action bomb. Bud and the other men of the 517th search all occupied buildings for additional bombs, but none were found. On November 4, 1944, First Battalion relieved Third Battalion at Sospel. Rumors began to spread that relief was soon to come to the 517th. With winter coming they thought for sure they would be able to get out of the Maritime Alps. The rumors became true when the 14th Armored Infantry and 19th Armored Divisions fresh from the boats relieved them at Marseilles. The 517th was ordered to La Colle-sur-Loup, six miles west of Nice. This would become a staging area pending their next combat mission. The 517th marched from Sospel to La Colle-sur-Loup and arrived three days later on November 18, 1944. Archer (1985) stated, "Pup tent camps were set up and efforts made to catch up on administrations. Inspections were held and clothing and equipment issued to replace the worn out gear they had. Duffle bags that were left in Italy had finally caught up to the men. Five hundred new replacements joined the 517th."

40 and 8 box car that Bud and other paratroopers were transported to the Battle of the Bulge.

On December 1, 1944, the 517th was ordered to join the XVIII Airborne Corps and directed to hold up in the town of Soissons in Northern France. The movement over the five hundred miles from La Colle-sur-Loup in southern France would be in train boxcars, better known as the forty and eights. This came from the length of the cars of forty hommes and eight ch eveaux. This meant 40 men or 8 horses. These train cars were the same type used the Germans to transport the Jews to the concentration camps. To move the men of the 517th would require three trains. One train assigned for each battalion. Thirty-five men were put into each forty and eight and shipped northward. Bud and the others only had C rations for food. Sleeping was done anywhere a soldier could find room. Some men made hammocks by slinging ponchos and shelter halves from the roof of the boxcar. For the next few days there would be no washing or shaving and the boxcars began to stink. To make matters worse the French railway workers would only take the train within their own localities. The train would stop to change crews every fifty miles or so. There were no scheduled stops so when a man had to use the toilet or relieve himself he had to jump off of the train when it slowed down. This was fine for the first two trains those men could jump back on another train. However, the men on the third had nowhere to go and found themselves in the middle of France. Some decided to take a few days R&R (rest and relaxation) before reporting back to the unit. On December 9 and 10, 1944, the trains arrived.

The 517th had been in combat for over **one hundred (100) days**. Everyone was hoping the war was now going to be over. Rumors had it that the war would be over by Christmas, but no such luck. Hitler would launch one last desperate campaign to try and stop the American forces. It would become to be known as the Battle of the Bulge. As the 517th arrived in Soissons, France it was cold, and then it began to rain. The unit stayed there twelve days rearming, refitting and obtaining new soldiers to take the place of the combat dead and wounded. With the war practically won the 517th returned to garrison duties. A new training order directed the men to begin close order drill (marching), tactical exercises, and calisthenics. The men began to believe they might have seen their last days of combat. While in Soissons a message was put out for all men who received a court martial in southern France were to report for additional extra duty in Soissons. Bud and Sug Lawrence reported to their Company Commander, Captain (CPT) Ehly. CPT Ehly told these two men they were once again on extra duty. How could this happen, they had performed their thirty days of extra duty while in southern France? It was not fair, but they found themselves once again performing dirty jobs no one wanted to do. They thought they would never get off extra duty when the word came down the Germans had launched a winter offensive. The 517th was quickly put back into action. Bud was happy he was off extra duty, but had no idea how tough it was going to be fighting in Belgium and Germany.

Letter to Mom from Harland "Bud" Curtis
Northern France, later to be know as Soissons close to Reims
Thursday, November 23, 1944
Thanksgiving Day

Dear Mom,

I have another money order I am enclosing in this letter. I sent the receipt for it in a letter to Jill. I figure one of them if not both should surely get there.

I haven't had any letters from you for over a week, but I guess it is because there are so many packages in the mail right now. That's right I got a package from you last Saturday with the cookies, pineapple, baked beans, boy it didn't take long to finish that off. I sure hope I get another one soon. How about in one of the packages you send, put in some of those assorted cookies.

I got a letter from Jill yesterday, and two from Jerry this week and one from Onieda Martin (or Marlin) a while back. I haven't heard from Dean Hildreth in months, so I don't know much about what he is doing either. He never did write very much. I don't have mail coming in regular like from my buddies, but often enough to know they are still alive. I heard from Frank Hill (Frank was the school bully who Bud and others were afraid of when they were children. Bud explained, "After the war in 1945, Dean Hildreth and I returned from fighting as combat paratroopers in Europe and one night Dean and Frank were together when some guy got out of line and Dean beat the stuffing's out of him. Frank was shocked. Later Frank and I were together when some guy smarted off to me. I beat this fellow within an inch of his life." Once again Frank was surprised and couldn't understand how these two boys he use to bully around when we were kids had become so tough. The answer was Paratrooper combat experience.) *Garth Boyce* (Bud's closest friend), *Garry Houston* (a neighborhood friend), *Dean Taylor* (a friend from church), *and others usually about every four or six weeks. Heck, I hardly ever write to anyone myself except you and Jill. I like to get letters, but I really don't care much for writing them myself. Every time I decide to catch up on back letters I usually end up by just writing you and Jill and saying to heck with the rest. I try to keep up with Jerry's letters though because I really do like the way she writes and I think it is nice of her to do it. Both of her last letters have been about 12 pages long and when I get finished reading them, I know what's going on around that part of the world.*

I hope Dad got that journeyman job he was talking about and I hope Bert is really on his way home for sure now. It would sure be swell if he could be home for Christmas, but as for me, being away from home at Christmas has gotten to be a regular habit, so it will be kind of nice sometime when I do stop adventuring around and settle down for a while in one place for a change.

Oh yes, today is Thanksgiving isn't it? We have a regular field kitchen now, and from what they tell me, I gather we are going to have a pretty good chow today with turkey and all the trimmings. It seems funny after so long of cooking your own food, to have someone else do it now, and sweating out a chow line again.

In one of your letters you said Irene called and was coming out, but I haven't had any mail since then so I still don't how things turned out.

I got that package of Esquire (magazines) and have been getting the books and magazines you have been sending quite regular. Thanks millions for all of them and I'll appreciate it if you just keep on sending them and packages of cookies and stuff like you have been.

Well the mail hasn't come in yet today, so maybe I'll get some mail from you. Hope so. I'll write again soon,

Love,   Bud

Letter to Folks from Harland "Bud" Curtis

Northern France
Saturday December 2, 1944

Dear Folks,

Gosh, I owe you all a letter now. I got one from Dad and you the other day, and just a few minutes ago I got a letter from Bert (Bud's brother) *telling me his experiences in the South Pacific. I'm sure happy he finally got home again safe and sound and I truly hope something will happen that he won't have to ever ship out that way again. I don't like to have him running around out there.*

I've got lots of things to thank you all for, so I'll try my best. I'll start by thanking you for the three packages that have come in the late two days, and also my watch finally arrived and it sure is a honey. It's really a classy looking watch and is keeping perfect time. I'm sure glad I have it. The packages I got had the addresses of "Hi Douglas" – Stienickerts and the one with Bert and Lorrain's address in Redondo Beach. The chocolates you sent were really delicious. I didn't know you could still get candy like that. I'm saving this strawberry jam and honey as I have a pretty good idea that I'll enjoy it more in a few days. Thanks millions for everything. These packages are coming through swell, and I guess I have been getting my share of them.

I got a package from Jill a few days ago and this is some stationary she put in and her typing on the envelope. She sent me some fur lined gloves and are they ever warm. She is sure nice to me I think.

Well, so this time you finally made connections with Irene (a girl Bud met while on a pass to New York City. Bud said she was a little older and acted like a big sister to him). *I'm sure glad that you liked her. I always thought she was a pretty nice girl, and she was sure swell to me in New York. She walked right off her job in the middle of the day once to come meet me at the Pennsylvania Train Station, because she was afraid I would get lost.*

I'm sure she enjoyed the visit with you and I do hope she can come down for the weekend sometime (Irene stayed at Bud's folks house while she visited southern California).

I wonder what Jill would say if she knew. Maybe she wouldn't care. Heck, Irene is just a real nice friend and I don't put any girl above Jill. I haven't met any girl that could change the way I feel about that Blond scatterbrain, so she hasn't got anything to worry about and I guess she knows that. I've sure told her enough times anyway, but I'm not sure it would be the best thing if they both happened to come out to the house at the same time. Women will be women you know, and far be it from me to ever try and figure out what they will do next, so maybe I better leave well enough alone.

Not much has happened around here to speak of. We still are having chow from a field kitchen (containers of food prepared in the rear and taken out to the men in the field, then served to them) *and there is a picture show about every night. Tomorrow we are going to have memorial serves for the fellows that were killed* (Soldiers from the 517th PRCT killed in December 1944, were 3 officers and 21 enlisted men. During the entire time in combat from Italy to Germany, June 1944 to February 1945, a total of 15 officers and 202 enlisted men were killed).

It is sure swell to hear Dad is the foreman of the department now, and can quit that night job at the shipyards (Bud's Dad, Bert senior, worked for the city of Long Beach as an electrician and also worked at the Naval Shipyard in Long Beach part-time as an electrician wiring ships at night). *Do you remember where Eddie Hunter use to live on 3rd Ave, and Atlantic Boulevard, at the Rose Court Apartments? I think it was Apt. H. If you ever happen to get around that way drop in and see if his folks are still there. I would sure like to get Eddie's address. I haven't heard from him for over a year now* (Before Bud was in the Army he, Garth, and Eddie left Long Beach in 1942 to go and work in San Francisco, Garth was the guy who came up with the idea of going to San Francisco. After living in San Francisco for 6 months, Bud and Garth went to Salt Lake City, UT, and Eddie returned to Long Beach).

I don't get many letters from Jill anymore. I hope she isn't sick. I imagine she is pretty busy with school and all, but I sure miss her letters (Jill was a senior at Wilson High School in 1944).

I got another letter from Phyllis (Proty) and Jerry (the girl from Sears) the other day and from Roger (Bud's high school friend) (Phyllis was a girl Garth introduced to Bud and they dated a few times).

Thanks again for the swell packages and I hope to get some more of your cookies soon and that's a help if anyone wants to know, so send all the things you can. I'm always happy to receive them. Tell everyone hello and,

Lots of love,
Bud

Letter to Mom from Harland "Bud" Curtis

Northern France, later to be known as Soissons close to Reims
Monday, December 11, 1944

Dear Mom,

Finally got some mail in today, and I got a letter from you and one from Jill with that picture I asked for. I'll bet it does seem nice to have Bert home and taking his motorcycle apart and getting all greasy. There is so darn little I can say, that I'm afraid I will have to make this letter do for everyone. Tell Bert I'm sure happy he made it home okay, and I hope he won't have to ship out for a while.

As you can see from the heading on top that I am now in Northern France, and I also have a new A.P.O. (Army Post Office). Which is "109", so remember it when you write next. I can't tell you anything about how I got here or when I left or arrived, but without giving away any valuable information I think it is safe to say – "Gee, but it is cold."

You know how the sky is there in Long Beach about 5 in the afternoon, well it is that way here all the time. Dark and dreary all day long so I can't take any pictures at all. I sure hope that Irene comes down for the weekend like you said she might. If you see her or Jerry (was a girl who worked at Sears at 5th and Long Beach Blvd. Bud's mother introduced them through the mail and he never met her until after the war, Jerry later married Bob Pack a medical doctor), *tell them I'll write as soon as I can, maybe tonight. I haven't had much chance to write for a while or get mail either, but I think things will get going regular now. I'm sure glad you like Irene as I think she is a pretty nice girl myself. If the sun ever comes out for a while she is there be sure and take some pictures and send them to me. Also if you can, take some more pictures of that scatter brain blonde and send them this way. In fact take as many pictures as you can of everyone as I sure do like to get them.*

If you get in the mood to send another package, I would sure like some more honey, and strawberry jam, and anything else you want to send. It sure tasted swell. The watch is keeping perfect time. It sure is a dandy, and I'm very happy to have it. Thanks a lot for everything.

Just two more days till my birthday. Gosh but time has gone by fast, and yet so slow. I hope I'm home this time next year. It is about time for chow and I'm about starved. Oh yes, the big event of the day was I took a super wonderful hot shower today, and I feel halfway clean for the first time in quite a few months. Boy, what I wouldn't give to step into that shower at home and stay there for half the day. How about some hair oil in one of those packages!

Well, I guess I had better start closing this letter and get ready for chow and I think there is going to be a U.S.O (United Services Organization) show tonight and I'm going to do my best to see it.

Keep writing often and give my love to everyone.

A' bientot and Monaime a tout or just so long and love to all.

Bud

Letter to Mom from Harland "Bud" Curtis

Northern France
Wednesday, December 13, 1944, Bud's 20th birthday

Dear Mom,

Got your letter this evening that you said hoped would get to me on my birthday, and it did exactly. I haven't much to say and it is pretty late at night anyway. I just want to send this money order. I sure hope you are getting all of them and the allotment.

It would be nice if you could find a lapel watch that Jill would like, but I guess they are hard to find. I don't care how much it costs, whatever you do get her; just as long as it is something nice and will make her happy. The only thing I want for Christmas as any other time is to have her happy. She is about the only thing that keeps me from getting discouraged and loosing hope of ever getting back so take good care of her for me.

Tell Bert and Pop hello and also Lorrain and Harley and everyone.

Lots of Love,

Bud

TOP SECRET

HEADQUARTERS 517TH PARACHUTE INFANTRY CT
APO 758, U. S. ARMY

Copy #27

11 August 1944

1. The following general considerations should be noted in the operation now being undertaken.

    a. For most of us this is our first combat jump. Consequently some may be a little apprehensive. Remember that the advantage is with the attacker, as the enemy does not know exactly how or when he is going to be struck. Particularly in an airborne operation in which we land in his rear areas where his CP's, lines of communication and supply echelons are set-up, will our activities give him grave concern. The enemy consequently will be a lot more apprehensive than we are. That the enemy will react to our landing by movement of forces towards our area should be expected within a few hours. However by that time we expect to be pretty well set for him and deal out a lot of punishment. It must be remembered also that he will be engaged at many other points by other airborne units and the large scale attack by the amphibious landing. There can be no doubt of the success of this operation if we use our heads and keep our confidence, work quickly but smoothly and act aggressively using good tactics and security measures.

    2. As it will be dark when we first land it will be difficult to see what is going on around us. Therefore, we must not start firing promiscuously at any thing that moves. You must be certain it is the enemy before you fire. After the first unit lands many of our men will be moving around the area, some already in assembly or moving to assembly positions and others recovering bundles. Don't fire first and find it is one of your own men later. Weapons, on landing, should be loaded and locked, and fired only on orders of an officer or in case of emergency. The sound of enemy weapons is known to you and should disclose the location of enemy forces if present. Enemy flares may be fired around the area to give the appearance of enemy strength and to cause us to be alarmed. Remember that more casualties can be caused by some of our own men getting trigger happy than from enemy fire. It is possible that your stick may land some distance from the DZ. Your action in this case should be considered. The general idea is to move to your Battalion assembly area and if the Battalion has already left for its objective join them there. If this is impossible join up with friendly troops preferably of this unit and assist them in accomplishing their mission. In either case do as much damage as possible on the way, for example a staff car or a truck load of Germans may be driving along the road unaware of your presence in the area. In case no officers are present with your group the senior NCO should be prepared to take over. All men will be issued instructions regarding the terrain, objectives, important towns, streams etc., and also a map in the escape kit. Remember that a few men can create a hell of a lot of trouble if they happen to be in the right place.

(O V E R)

TOP SECRET

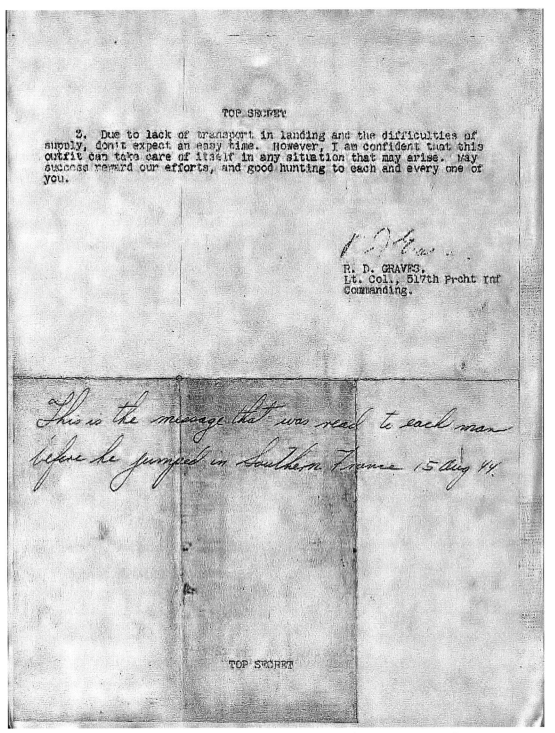

Colonel Grave's original Letter to the Men of the 517th before their combat jump

Chapter 13

THE BATTLE OF THE BULGE - BELGIUM AND THE FINAL BATTLE - GERMANY

(The 517th PRCT was attached to the 82nd Airborne)

The picture of the paratrooper at the beginning of this chapter reflected the attitude of paratroopers of the day.

The text reads:

Dec. 23, 1944 - "Battle of the Bulge" - An entire U.S. armored division was retreating from the Germans in the Ardennes forest when a sergeant in a tank destroyer spotted an American digging a foxhole. The GI, PFC Martin, 325th Glider Infantry Regiment, looked up and asked, "Are you looking for a safe place?" "Yeah" answered the tanker. "Well, buddy," he drawled, "just pull your vehicle behind me...

"I'm the 82nd Airborne, and this is as far as the bastards are going."

The poster is a photograph of a dirty, scrappy, tough paratrooper, PFC Vernon Haught, of the 325th Glider Infantry Regiment, marching in the dead of that cold, snowy winter with a rucksack on his back. Going to reinforce the retreating American forces in Belgium. His expression leaves no doubt about his determination. He is moving out to go toe-to-toe with the enemy in Belgium. As you look at the poster, it strikes you that nowhere in this photograph do you see a parachute. And you and I both know there doesn't have to be one -- you simply know from the look: he's Airborne. Imagine, an Airborne PFC telling a sergeant in a tank to follow him. That is the tenacity of an Airborne trooper. The best America could offer to defend the country.

While the 517th was held up in Soissons France rumors began to go around that they would be part of a new airborne assault across the Rhine River sometime in the Spring of 44. The idea of the war ending was now starting to fade away. Then the official word came down that all airborne units were now reassigned to the XVIII Airborne Corps. This included the 82nd, 101st, and 17th Airborne Divisions, along with separate Regimental Combat Teams such as the 517th. Other units were still in combat, but all paratroopers were pulled out and were being organized for one last jump into Germany, and it would all be over. Hitler, however had different plans. On December 15 and 16, 1944, the German Army launched its last great offensive of World War II, catching the Allied forces completely by surprise by striking with three Armies against a weak American position in the Ardennes region of Belgium and Luxembourg. Hitler wanted to seize the port of Antwerp and drive a wedge between the Americans and British. The famed 101st Airborne Division was at Bastogne at Christmas time when all American forces were cut off and surrounded by the Germans. On December 22, 1944, the German commander sent Brigadier General McAuliffe a letter demanding his surrender. On the next page is the letter he sent back to the German Commander. Brigadier General McAuliffe wrote this letter to his men at Christmas. News spread quickly and all airborne units were assigned to defensive position in the areas of Bastogne and St. Vith.

Letter of Brigadier General McAuliffe dated 24 December 1944:

"Merry Christmas"
Headquarters 101st Airborne Division
Office of the Division Commander - 24 December 1944

What's Merry about all this, you ask? We are fighting - it's cold, we are not home. All true but what has the proud Eagle Division accomplished with its worthy comrades of the 10th Armored Division, the 705th Destroyer Battalion and all the rest? Just this: We have stopped cold everything that has been thrown at us from the North, East, South and West. We have identifications from four German Panzer Divisions, two German Infantry Divisions and one German Parachute Division. These units spearheading the last desperate German lunge, were headed straight West for key points when the Eagle Division was hurriedly ordered to stem the advance. How effectively this was done will be written in History; not alone in our Division's glorious History but in World History. The Germans actually did surround us, their radios blared our doom. Their Commander demanded our surrender in the following impudent arrogance.

December 22nd 1944
To the U.S.A. Commander of the encircled town of Bastogne,

The fortune of war is changing, this time the U.S.A. forces in and near Bastogne have been encircled by strong German armored units. More German armored units have crossed the river Ourthe near Ortheuville, have taken Marche and reached St. Hubert by passing through HomprÃ-Sibret-Tillet. Libramont is in German hands. There is only one possibility to save the encircled U.S.A. Troops from total annihilation: that is honorable surrender of the encircled town. In order to think it over a term of two hours will be granted beginning with the presentation of this note. If this proposal should be rejected, one German Artillery Corps and six heavy A.A. Battalions are ready to annihilate the U.S.A. Troops in and near Bastogne. The order for firing will be given immediately after this two hour's term.

All the serious civilian losses caused by this Artillery fire would not correspond with the well-known American humanity.

The German Commander

The German Commander received the following reply:
22 December 1944

"To the German Commander: N U T S !

The American Commander

Allied Troops are counterattacking in Force. We continue to hold Bastogne. By holding Bastogne, we assure the success of the Allied Armies. We know that our Division Commander, General Taylor, will say: " Well done!" We are giving our Country and our loved ones at home a worthy Christmas present and being privileged to take part in this gallant feat of arms are truly making for ourselves a Merry Christmas.

(signed)
McAuliffe,
Commanding.

At 1100 hours on December 21, 1944, the 517th received it's orders to move out through Namur to Werbormont Belgium. Warnings were give to stay clear of Bastogne since it was cut off and surrounded. The men were quickly given whatever cold weather clothing that could be found. Bud only had an overcoat, gloves, and one pair of long underwear. Fortunately for him he had the knitted wool cap with extended earflaps and muffler that Jill had made for him during Tennessee maneuvers. Bud described that when the other men saw him with that muffler and warm cap they were envious, and upset he had something they didn't have. In fact some were just hoping he would be killed so they could take that warm muffler and cap.

At 1800 hours, Bud and the first Battalion rolled out, followed within a few hours by the 2nd and 3rd battalions. Bud describes the ride in the 2 ½ ton trucks (commonly called duce and half) was long, miserable sleepless night, and cold. Sleet, snow and freezing rain fell continually. It was the worst winter in twenty years. The men were not prepared for this harsh winter weather and did not have adequate winter clothing, but they had the Tennessee maneuvers under their belts. Bud described that the weather in Belgium as very cold, usually twenty below zero at night, but in retrospect the Tennessee maneuvers were worse because it rained and snow and he was soaking wet most of the time.

When the 517th arrived at Werbormont on the morning of December 22, 1944, the First Battalion was reassigned to General Rose's 30th Division at Manhay. Lieutenant Colonel Boyle, the Battalion Commander did not know where Manhay was, but found Manhay without a map by asking a truck driver, "which way to Manhay?" When the battalion arrived General Rose told the First Battalion to report to town of Soy and prepare to attack. The First Battalion arrived at 1600 hours and by 1715 hours the battalion began the attack from Soy to the town of Hotton. As they moved through Soy they received light artillery fire. Bud was use to this noise and harassment as he had survived the tunnels in Col de Braus France, but it didn't make it any easier. The Germans held the high ground, and as the First Battalion maneuvered through Soy it became difficult to advance. The Germans occupied a hill between the towns of Soy and Hotton and just as soon as the first battalion cleared the town of Soy they came under machine gun fire from the hill. Then more machine gun fire coming from the south of the road intersection. This made the battalion's left flank totally vulnerable. The light of day quickly faded as night came on, and the first battalion held its position. Then "B" Company was sent in to dislodge the enemy.

Sometime during the fight for the town of Soy, PFC Melvin Biddle went above and beyond the call of duty during the fight. On March 17, 2006, it was my pleasure (L. Vaughn Curtis) to travel to Anderson Indiana and visit with Mr. Melvin Biddle. I had told Mr. Biddle my father Bud Curtis was in the First Battalion. Mr. Biddle invited me to visit with him. Mr. Biddle told me he was a member of "B" Company, First Battalion. Mr. Biddle told me he was part of a patrol sent out to rescue a company of cooks and clerks that had been encircled by the Germans near the town of Soy on December 23rd. Two lead scouts were injured when one of the guys stepped on a mine. Mr. Biddle said his Commanding Officer (C.O.) told him to get out front and act as scout for the company. So crawling through the snow PFC Biddle came across a German outpost. There he killed three Germans soldiers and then moved forward until he saw a German "Machine Gun Nest." PFC Biddle threw a hand grenade and took out the machine gun nest. Waving for the company to come forward he advanced further destroying two more machine gun nests. Mr. Biddle said, "My C.O. arrived and ordered me to go out and find a prisoner. Hiding in the snow a German patrol passed me. Then I saw a German officer and I came out from my hiding position and pointed my rifle at him trying to take him prisoner. The German officer would not surrender, pulled out his Luger handgun, firing it wildly, and ran off." PFC Biddle returned to his company and took up a position in the snow and tried to keep from freezing. Mr. Biddle continued with his story relating how bitterly cold it was. He expressed how he worried all night long about his fingers freezing and not be able to pull the trigger on his rifle. Surviving the night's cold, the C.O. sent PFC Biddle out again to find a prisoner. This time he saw a thirteen man German patrol running through the snow. He shot and killed all thirteen soldiers. As he moved forward he saw a boy in a German uniform tied to a tree with rifles and hand grenades at his feet. Other 517th troopers told him to shoot the boy German soldier, but PFC Biddle took him prisoner. One week later PFC Biddle was hit in the neck with shrapnel just missing his jugular vein. He was evacuated to England where he was treated. After his wound healed he was heading back on a train to rejoin the 517th when he read in the "Stars and Strips" newspaper that he was being awarded the Medal of Honor. After the war on October 12, 1945, PFC Biddle was awarded the Medal of Honor by President Harry S. Truman. Mr. Biddle told me, "In my official picture receiving the Medal of Honor from President Truman they gave me Corporal stripes to put on my uniform, but I was really a PFC". According to Del Calzo and Collier (2003), a total of 2400 Medal of Honors had been awarded to U.S. Army Soldiers. Indeed few of those were Army paratroopers. Melvin Biddle's tenacity as one of the toughest of the tough not only brought him distinction, but also brought great honor to the paratrooper forces.

Melvin Biddle, Company B, 517th PRCT receiving the Medal of Honor from President Harry S. Truman.

Now back to Belgium 1944, at Soy and Hotton. At day break the First Battalion decided to do an envelopment maneuver by going north of the town of Hotton. The fight was fierce. The Germans were not giving up an inch of ground. They knew this was their last chance at winning the war and forcing the allies to a peace table. The fuehrer had demanded that each German lay down his life for the fatherland.

The First Battalion didn't know, but the unit they were fighting was the 116th Panzer Division. This division was the best Germany had to offer and these men were ready to die for their mad dictator and the fatherland. The 517th would do all it could to accommodate these German soldiers. An hour after midnight Company A with four tanks and six half tracks broke through the German lines. By 0630 hours Lieutenant Colonel Boyle, and the rest of the First Battalion entered in to the town of Hotton. It had been a rough battle, the Germans were not easy foes, and just when Bud thought maybe the Germans would retreat; the Nazi Army ordered in the 560th Volksgrenadier Division to relieve the 116th Panzers. This meant more fighting with a fresh German Division. The exchange made no difference to the First Battalion, they were not moving and the Germans were unable to retake the town and forced the 560th Volksgrenadier Divison to withdraw to the south. The First Battalion's attack of December 22-24 caused the Germans to retreat to La Roumiere Ol Fagne, a mile south of the town of Soy. On December 24th, the 290th Infantry Regiment was ordered to attack La Roumiere Ol Fagne. The attack was to occur at 2300 hours. Much confusion ensued and as they attacked the Germans. The Germany Army found additional divisions to counterattack. From the hills of La Roumierre the German 1129th Regiment watched with amazement at the men of the U.S. Army's 290th Infantry Regiment moved into open fields. The Germans either thought these men were extremely brave or stupid. They fired mortars, artillery, and machine guns. These American were cut to pieces. The causalities were catastrophic. Most all officers were killed and by 0800 the 290th was pinned down in the snow. The 290th Regimental Commander ordered the reserve 3rd battalion into the same area. They became pinned down with many losses of life. Colonel Howze, the 290th Regimental Commander, order Lieutenant Colonel Boyle, First Battalion, 517th to get his battalion together and secure La Roumiere, Belgium. An hour before dark the First Battalion moved south from the town of Soy west across the wooded hills of the Melines la Fret. Crossing steep terrain they met resistance on the eastern slopes of La Roumiere from the German 1129th Grenadiers Division. The battalion encountered stiff resistance from the Germans and Lieutenant Colonel Boyle called in artillery support fires that drove the Germans out of La Roumiere and forced them to withdraw to the south.

Colonel Howze of the 290th Regiment placed Lieutenant Colonel Boyle in command of all forces on La Roumiere, to include all units of the 290th. On Christmas morning the First Battalion was relieved and by 1500 hours the weary battalion left La Roumiere on trucks and rejoined the 517th regimental control at Soy, Belgium. It had been a rough four days of combat. Once back in Soy the battalion found out that the town of Manhay had fallen to the 2nd SS Panzer Division of Christmas Eve. After the loss of the Fraiture crossroads on December 23, 1944, General Ridgway ordered the 3rd and 7th Armored Divisions along with the 82nd Airborne Division to move toward Manhay and retake the town at all costs. At noon on Christmas the 2nd SS Panzers were putting up a decisive defense, and was holding off the 3rd,7th, and 82nd divisions. By 1400 hours General Ridgway ordered the 517th to attach one battalion to the 7th Armored Division. Third Battalion was committed with the First Battalion moving on December 26, 1944, 12 miles from Ferrieres into an assembly position in Harre five miles north of Manhay where they could block a counterattack. The Third Battalion took Manhay, and on December 26 and 27th the enemy advance had been halted. For nine long cold miserable days the 517th held the southern flank of the XVIII Airborne Corps' line. Enemy penetration had been stopped at towns of Soy, Hotton and Manhay.

The German SS Panzer Division had been badly beaten and the 560th Volks Grenadier Division was now crippled and unable to fight. Hilter's last great military offensive had been crushed. General Ridgway would endorse a letter of commendation to the men of the 517th. Now finally on December 29, 1944, Bud was able to write his mother and let her know he was alive and well. He would not write to her about the horrors of this campaign, and he would never talk to them about it ever.

```
                    R E S T R I C T E D

BASIC:  Ltr, Hq 36th Armd Inf Regt, SUB: "Commendation of 1st Bn, 517
            Parachute Regiment", 28 December 44.


201.22                              2d Ind.
(23 December 44)
HEADQUARTERS VII CORPS, APO 307, U. S. Army,

TO:  Commanding Officer, 517th Parachute Regiment.

THRU:  Commanding General, XVIII Corps, APO 109, U. S. Army.

        It is with pleasure that I transmit this tribute to the superb
fighting qualities of 1st Battalion, 517th Parachute Regiment and their
commanders, Lieutenant Colonel W. J. Boyle and Major D. W. Fraser.

                                /s/  J. Lawton Collins
                                /t/  J. LAWTON COLLINS,
                                     Major General, U. S. Army
                                            Commanding.


201.22 (CG)              3rd Ind                          MBR/ajv
HEADQUARTERS, XVIII CORPS (AIRBORNE), APO 109, U. S. Army, 16 January
1945.

To:  Commanding Officer, 517th Regimental Combat Team, U. S. Army.

        The Corps Commander takes pleasure in forwarding this commendation.


                                /s/  M. B. Ridgway
                                /t/  M. B. RIDGWAY,
                                     Major General, U. S. Army
                                            Commanding.


                    R E S T R I C T E D
```

General Ridgway's endorsed letter of commendation to the men of the 517th.

EXECUTIVE ORDER NUMBER 9396

PRESIDENTIAL DISTINGUISHED UNIT CITATION

FIRST BATTALION, 517th PARACHUTE INFANTRY REGIMENT

"The 1st Battalion, 517th Parachute Infantry, is cited for outstanding performance of duty in action against the enemy during the period 22 through 26 December 1944. Attached to the 3rd Armored Division at the height of the German counteroffensive, committed to battle in the vicinity of Soy, Belgium, at 1700, 22 December 1944, the 1st Battalion fought continuously until 1700, 24 December 1944, to contain the German attacking forces and regain the initiative. In bitter cold, and against numerically superior enemy forces, the 1st Battalion accomplished all assigned missions in fighting its way from Soy to Hotton, Belgium, checking and displacing the enemy forces driving towards Liege and Namur. This battalion captured the high ground controlling the road nets at Haid Hits and Hotton, cleared the enemy forces in its sector, relieved the beleaguered American forces at Hotton, and established a main line of resistance on a 6,000 yard front from Soy to Hatton, from which an attack was launched to the south by a fresh infantry regiment. At 1200, on 25 December 1944, when this regiment failed to accomplish its mission, the 1st Battalion, 517th Parachute Infantry, with less than a day's rest was further committed, and captured, with a handful of remaining men, the hill objective La Roumiere 01 Fange, from which a reinforced battalion had been repulsed with severe casualties. Through repeated displays of individual and collective gallantry, the officers and men of this battalion accomplished with distinction every mission assigned their commander and gained the appreciative admiration of all adjacent units with whom they fought. Suffering 157 casualties, the 1st Battalion accounted for 210 enemy dead and 18 captured. This battalion exhibited a high sense of duty and will to fight and, through the resulting achievements, it reflected great credit on its parent unit and the armed forces of the United States."

(General Orders 50, Headquarters, United States Forces, European Theater, 27 Dec 45)

MEN OF THE 1/517th PIR KILLED (21-26 Dec)

Lt. HARRY D. ALLINGHAM
Lt. ROLAND A. BEAUDOIN
Sgt. STANLEY S. BROWN
Sgt. CHARLES A. CRITCHLOW
Sgt. WILLIAM M. DELANEY
Sgt. DAVID A. RIVERS

Cpl. EDWARD J. LANG
Cpl. FRANK L. TIMINSKI
PFC JOHN MITCHELL
Pvt. LEONARD M. FANCHER
Pvt. LEDLIE R. PACE

The First Battalion, 517th PRCT received the Presidential Unit Citation

In Bud's next letter dated December 29, 1944, he related that the 517th PRCT had taken and secured the towns of Soy and Hotten Belgium. Bud thought he remembered the 517th turned control of these two towns over to the 106h Infantry Division, and on Christmas day the 106th Division had lost the ground given them by the 517th. Bud thought the 106th Infantry Division was pushed back by the Germans at the towns of Soy and Hotton. Because the battle was unclear to Bud, his son L. Vaughn Curtis asked the question on "Mailcall", the 517th PRCT Association email site in May 2004. On May 27, 2004 an email message was sent out to me from then Lieutenant Colonel Boyle, 1st, Battalion Commander, 517th . He straightened out the story as to what really happened. Colonel Boyle related the following on mailcall #700. He said,

> "This is to Lory Curtis, I Bill Boyle remember, but would change a few details. It was troops from a regiment of the 75th Infantry Division. Not the 106th Infantry Division. There Germans had not taken back Soy and Hotton, but two battalion of this regiment (517th) were unable to take the hill, Laremoulier. Colonel Howze of 3rd Armored Division ordered me to take the hill. At the time C Company was off on another mission. B Company was on a line from just south of Hotton to just south of Quatre Bras. I gave A Company the job of going down a streambed to attack the hill from the flank. One platoon of C Company that was not on that C Company mission was in reserve. A Company swept the hill although having been fired on by one battalion of the 75th Infantry Division troops. Just before dark I was ordered to take command of all our troops in the area and organize for a defense. I gave A Company an area to defend and then placed elements of about seven companies in position. As daylight came I heard an observer incorrectly directing fire and tried to correct it. It came in directly on part of A Company. It turned out to be from the cannon company of the regiment of the 75th Infantry Division. Yes, I raised hell about that as well as with the only battalion commander of that unit that I could locate. Our versions vary somewhat, but after all it is almost 60 years ago, and we saw if from different view points.

Bud remembered that artillery attack very well, (he told his mother about it after the war was over in his letter dated May 29, 1945) and has never forgot it all of these years later. He said, "On December 26th, I was stringing field telephone wire up to LTC Boyle, the Battalion Commander while conducting an offensive operation to retake Soy and Hotton. I had spliced the wire many times that day as German artillery rounds kept blowing up the lines. At about noontime on December 26, 1944, I was stringing wire for the field telephone for Lieutenant Colonel Boyle, the First Battalion Commander, when friendly fire from the some Artillery Battalion of the 75th Infantry Division came screaming in. Apparently grid coordinates for the intended rounds were landing short. LTC Boyle and I hit the ground as the shell exploded with most of the shrapnel going upward and not downward."

Bud then continued, "Close by were two men. One man was the forwarded observer with a radio on his back from that Artillery Battalion standing by the foxhole. The other man was a 517th paratrooper in the foxhole. When the rounds came in the man with the radio strapped to his back dove into the foxhole on top of the other man. The artillery shell exploding killing the radioman and his body covered the other man lower in the foxhole. This man started screaming and lost all control about the radioman's death." Sixty years later Bud could still hear his screams. After the artillery rounds stopped coming in, Colonel Boyle, and Bud were shook up badly, but not hurt. LTC Boyle asked Bud, "Curtis are you still alive?" Bud checked himself and said, "Yes sir, I think I am." The Colonel said he had to go and get things organized but would be back. LTC Boyle did come back and kept reporting to Bud the condition of the battle and reassured Bud that everything was alright.

Men of the 1st Battalion, Major Don Fraser, PFC Bud Curtis, and Colonel Bill Boyle discuss issues of the battles at Soy and Hotten, Belgium at the 517th reunion in Washington, D.C., June 30, 2007.

517th PRCT Airborne troopers in a Jeep at the "Battle of the Bulge"

Major Fraser, Executive Officer, 1st Battalion was carrying Bud and another trooper in the back of his jeep. Major Fraser noticed Bud's hands were soaking wet. He gave Bud his warm dry gloves saying, "Curtis take these, I can get more, you can't." Major Fraser put Bud's cold wet gloves on the radiator of the jeep to dry out. Gloves and jeep courtesy of American Patrol Company, Owner Ray Melbrum.

Letter to Mom from Harland "Bud" Curtis
Stavelot Belgium, Thursday December 29, 1944, postmarked January 24, 1945

Dear Mom,

This is the first chance I've had to write you for sometime and I'm afraid for a while now my letters are going to be few and far between, so try to understand that it isn't my fault.

I'm not going to tell you anything about these past few days because you would only worry about me and I don't want you to do that. I'll tell you about it after it is all over, but for now all I'll say is that I'm up front fighting here in Belgium and it is plenty cold

Right now I'm a little ways back from the front lines (in Stavelot Belgium) *and I am in a big mansion just like you've seen in the picture shows. I am in one of the big rooms and I am warm and even have a radio sitting in front of me listening to a broadcast from England, so I am enjoying myself while I can. Yesterday I had on e of the fellows that use to live on a farm get me a whole canteen cup of milk out of a real live cow. Boy, it was good just like when I was back in Payson and Uncle Willis* (Bud's father's brother) *milked that old cow of Ma's* (Bud's paternal grandmother).

I got a lot of mail yesterday and will probably get some more today if we stay here. I got those pictures of you and Dad, Ma, and Grandmother. They sure turned out nice and clear. I got seven letters from Jill and one from Gary and one from Jerry. I sure was glad to get so many from Jill and I am glad she has gotten over that cold she had. I'm going to see if I can find some more ink. I did. The clipping you sent me was about our outfit, but I'm a long ways from there now. I came all the way across France in one of those 40 and 8 box cars (train) *you heard about in the last war. You won't find much about the 517th by looking for news in the paper because we are a "Combat Team" and wherever there is some place they need good men fast that is where we go. We have been attached to almost every Army over here at different times.*

The reason we don't get much rest is because we are a good outfit and all the time someone is needing us. There is nothing we can't do, but you people won't read about it because we aren't a big enough outfit to make front page headlines.

I was sure thinking a lot about you people at Christmas and wondering what you were doing and if maybe you were thinking about me that day while I was ducking bullets and shells, and if there would ever be a time when we could all be home for Christmas together again. I saw a lot of guys that day both ours and Krauts that will never see another Christmas, and all I could do was pray to God that it wasn't my time to go, and I guess it wasn't cause here I am. I'll never forget this Christmas as long as I do live, and you will never know how glad I am that this war is being fought in these countries and not in ours and I'm glad that you people there will never have to see some of the awful things I have, but I don't even want to think of them so I'll drop the subject.

I sure hope Bert was home for Christmas and Jill was happy with what ever you got her from me. Write soon and tell me everything and tell everyone hello from me and I'll write again as soon as I can.

I'm going to heat me up something to eat now so I'll say so long for now, and I hope some more mail comes in tonight. It is sure swell to get mail from home so keep it coming. Oh yes, I got two packages from you just before I left that last area (censorship would prevent Bud from telling his mother exactly where he was in case the Germans intercepted his letters home. Each of his letters were opened by Army mail personnel and screened before they were sent out) *to come up here in Belgium. Thanks a lot for them they sure tasted good. Believe it or not I ate that whole can of pineapple and fruit cake laying on the ground with machine gun bullets whizzing only about 6 inches over my head. I figured at least I was going to eat that instead of some darn Kraut, come what may. Send some more when you can and some cookies and candy. Lots of Love, Bud*

During the "Battle of the Bulge" Bud had been out most of the night splicing commo wire for the telephones. Sometime around 2:00 or 3:00 AM, Bud came stumbling into this farmhouse. Inside were many paratroopers from his unit who were sitting by the fireplace playing cards. A fire is roaring and Bud takes off his overcoat that is soaking wet and hangs it by the fireplace to dry. He walks upstairs and by chance finds an empty bed. Bud is exhausted and lays down his rifle and collapses on the bed. Within seconds he is sound to sleep. Bud remembers what happened next. He related that his friend Sug Lawrence jumped on his chest yelling, "Get up Curtis, the Germans are just outside." Bud was startled out of a deep sleep. He didn't know what to think. His first instinct was to grab his rifle, which he did. Then he looked out the window and could see German soldiers moving around the houses. Bud ran down stairs quickly, there at the door was a paratrooper saying come over here. As he looked out the door and when he thought the coast was clear he would tell the men to run to safety. Bud stood there until his turn. Then the words "Run." Bud took off running as fast as he could, crossing an open snowy field until he arrived with the others paratroopers. Just then Bud noticed he had forgotten his warm overcoat. He was grateful that he hadn't taken anything else off. The weather was bitter cold (Bud tells his mother about this in a letter dated February 19, 1945 in his postscript). In addition to this story Bud will always remember the day his battalion executive officer took care of him.

Sometime in early January 1945, while in Belgium during the Battle of the Bulge, Major Don Fraser, Executive Officer of the First Battalion was transporting Bud and some other soldiers to another location in his jeep. Bud had taken off his gloves because they were soaking wet. Major Fraser noticed that Bud was trying to keep his hands warm with no success. Major Fraser took off his gloves and told Bud, "Here take these gloves I can get more but you can't." Bud was very grateful and never forgot Major Fraser's kindness.

Since learning about the glove incident I wanted to know more about it. I posted an email message on the 517th mailcall site, asking if Major Fraser was out there and if he remembered the glove incident. Much to my surprise the following email was posted to the site on May 27, 2004:

In an email message, Jay Littlefield, Major Don Fraser grandson wrote this about the gloves: "This is Jay, the 20 year old grandson of Major Don W. Fraser. I showed him the story that your father remembered, and my grandfather immediately picked up on it. He told me to tell you that the jeep was from Rome, Italy, and he stole it from the British for Colonel Boyle. He said that it had no top or windshield. He also added, "Keogh was my driver, he could see in the dark like a cat." He also remembered the glove incident as well. He wanted me to tell you that he did get another pair of gloves from Bill Price, the S-4, and that the wet gloves were placed on the radiator to dry off. He was very happy to hear that someone else remembered that jeep and gloves, etc. He also added that Airborne units never got many vehicles, so whenever they would need something they would steal it. He said that they just had to paint over the numbers on the bumper and paint 517 instead. On this jeep they had to paint over the British emblem on the hood as well. Thank you very much for your email, my grandfather's eyes lit up when he read the letter. It was good to see that he has his memory so intact."

On July 19, 2006, at the 517th PRCT reunion in Portland, Oregon, Bud Curtis and Don Fraser met for the first time since the war. Both of these men remembered that day in Belgium when Major Fraser gave Private Curtis his gloves. Major Fraser related, "My gosh his fingers were frozen stiff and I knew he needed my gloves. I could always get more but he couldn't." To this day and all of his life Bud never forgot this kindness. Now these two men had time to talk about their lives, children, and grandchildren. It was a great reunion.

As 1944 ended, Bud and every other soldier that was in a combat zone were too busy fighting a war to reflect on the music of the day, but the rest of the world was listening to these hits of 1944: The Trolley Song, by Judy Garland, San Fernando Valley, I'll Be Seeing You, and Swinging on a Star, by Bing Crosby. Shoo Shoo Baby, by the Andrew Sisters. Is You Is, or Is You Ain't My Baby, by Louis Jordan. It Had to Be You, by Dick Haymes, My Heart Tells Me Should I believe My Heart, by Glen Gray It's Love, Love, Love, by Guy Lombardo. When the men did get to hear these tunes they knew what they had been fighting for.

Early on New Year's Day 1945, the Germans launched the Ardennes counterattack. The 517th was quickly attached to the 82nd Airborne Division, Bud and the First Battalion began marching from Harre, Belgium toward Trois Points (translated meaning three points). In Bud's January 2, 1945 letter he explained to his Dad how he marched for more than seven hours. Bud also told his Dad he had lost the wristwatch that was sent to him in a care package from his mother when he was in Soissons, France. As you read you will see what happed to his watch.

Letter to Dad from Harland "Bud" Curtis

Belgium
Tuesday January 2, 1945

Dear Dad,

I've got so many things to write about I hardly know where to begin. First off, I'll tell you I got another swell letter from you and you have all kinds of news in it. I hope you have gotten all the money orders I've sent home, and also I hope by now that check that I sent you has gotten there too. Let me know if they have it or not. I can trace the money order through the receipts but there isn't anything I can do about the check.

Now I have some good news and then some bad news and then some good news again to tell you. Well to start with it was yesterday, New Years day and I was just walking along the road with the rest of the boys headed up to the front lines. Well anyway there I was slogging through the snow when someone right behind me says "Hey Curtis", and I turned around and dog gone you could imagine the look on my face to be looking at Dean Hildreath. Boy what a thrill to see someone you know. He walked down the road with me for about 2 miles and we talked as we walked about a million and one things of where we'd been. I sure wish I could have stopped. For two days I was in a small town just only about a half mile away from him. He is in the 504th PCT now in service company taking life easy. He looked plenty good and healthy enough. It was just a coincidence how we happened to meet. He was staying in a house that we were passing by and he asked one of the guys what outfit it was passing and they told him and then he asked if any of them knew me and it happened that I was only about half a block up the road. I don't know when I'll see him again, but I told him to look up Willard Hill if he got the chance (Willard Hill was the Sunday School teacher in 1942 who was a paratrooper who gave Bud the idea of joining).

Well Dean had a turkey dinner waiting for him and he was walking farther away all the time, and I told him he had better start hitch hiking back so the last time I saw him he was heading back for that turkey. We kept on walking for 7 more hours after Dean left and boy you should have seen those slippery roads we were walking on. I liked to broke my neck a hundred times until I finally got here where I am now. Anyway the most roughest part of the way was just about two miles from here and somewhere along the way on one of the times I slipped and hit the ground. I lost that swell watch you sent me. Boy, I sure felt bad about it and I figured that some truck or jeep had buried it into the ground if someone hadn't picked it up, so I gave up hope of ever seeing it again, and dreaded having to tell you I lost it. Well that's the bad news but I have some good news again. I was checking a line down that same road I came up last night and a fellow I know came by in a jeep and asked me if I knew anybody who lost a wristwatch? Oh boy, I could hardly believe my ears. There the watch had laid all night and he just happened to spot it along side the road. The pin holding the strap had broken and now I've got it tucked way down in the pocket of the first pair of pants I have on so I'm not going to loose it again. It was practically a miracle I got it back and I'm sure happy I did. It keeps perfect time and really is a swell watch.

Now I'll tell you something where I'm at. Just picture yourself in a house something like Ma had in Utah, only here things are even more convenient. You just step out of one of the doors like as if you would be stepping into a garage from your house, but instead of having a car we got a big barn built right into the side of our house and 6 cows, 2 calves, 2 dogs, and a cat and some chickens, so I have been having milk, eggs and homemade butter and we killed 4 rabbits and really had a swell dinner. There is only one catch to it, I'm afraid I won't be here long, but I'm sure enjoying it while I can. I even have a big soft bed up stairs to sleep in and is it ever nice. We can't move around much in the day time as we are under enemy observation so we do most of our work at night.

I feel just like a civilian the way I've been living today, but by the time you get this letter I'll be someplace else.

I just came back to finish this letter. I ran off without saying anything and had a swell supper of milk, potatoes, crackers, and homemade butter. I could drink milk and eat this butter all night long. It is really delicious. We take the cream from the milk and shake it in a jar, and the next thing it is creamy butter white like ice cream. I guess I'm not telling an old farm hand like you anything new though am I? I don't know how I'd like to always be a farmer but I'm sure having the time of my life now. I haven't advanced far enough at this game to try milking a cow all by myself. I leave that up to some of these guys who use to live on a farm. I crawled up in the hay loft and pitched them down some hay to eat and gave them water, and even went as far as to scratch one cow's head. I don't know if it's head itched or not, but it seemed like a friendly thing to do for a cow after it gave so much milk. I'm not sure but I think it appreciated me scratching it's head cause it let out a nice loud moo and smiled. I guess I'm just a city boy at heart though.

I hope I get some mail soon telling me what you did on Christmas and New Years and I hope that Jill likes that charm bracelet.

I'll write again as soon as I can. Tell everyone hello for me. It snowed today, but has been warmer if this kind of weather could even be warm. Well I think I'll take advantage of that nice bed while I have it so I'll close shop for tonight. Write soon.

Love Bud

Bud described in August 2006, "It was a force march to Trois Point that was grueling and never ending. We marched all night in the cold snowy weather arriving just before daybreak. As my letter to my father said I slipped and fell down. I got back up and continued marching not noticing my watch was gone. Then as I said in the letter a friend of mine found the watch and gave it back to me. I couldn't believe getting it back. Once the war was over I let my older brother Bert borrow the watch as he was traveling east on a bus. My brother opened the bus window somewhere crossing Arizona because it was hot, and the wristwatch fell off of his arm never to be seen again. When my brother returned to California he told me he had lost the watch, but he gave me another watch to replace it. I still have and treasure that watch today."

In Bud's letter of January 2, 1945, he described having potatoes, milk eggs, homemade butter and cooked rabbit. Ever since Bud was a young boy, he loved to drink milk. Since arriving in Europe, and while in combat he had very few opportunities to drink milk. Other men in the unit enjoyed milk as much as Bud, and also found it difficult to find the opportunity to get a glass of milk.

Howard Hensleigh, a Lieutenant in Headquarters Company, Third Battalion of the 517th described what it was like to finally get a chance to drink milk while in combat. In an email message on the 517th email site in November 2005, he recounted his experiences during the Battle of the Bulge and milk:

"During our attack south of Stavelot, Belgium we had the high ground to take. On the second day of the advance, we came upon a farm family that had been able to stay put. The farmer had gathered the neighbor's cattle into his barn where he fed and milked the cows. At that time, we had not been able to get our hands on fresh milk for what seemed like an eternity. Our milk and eggs in Europe were dried.

Although the cooks, when we were out of combat, did their best, it is impossible to make dried stuff taste like the real thing. In combat, the cooks served as litter bearers, while we ate K rations when we could get them. As we went by in the attack, the farmer's daughter, a pretty ten (or so) year old, poured fresh milk into our canteen cups. To my dying day, I will always remember the thoughtfulness of that family and that pretty young Belgian girl. One of the neighborhood cattle did not make it due to an incoming mortar round. Lt. Col. Paxton shot the animal in the head with his forty-five as we ran by in the attack. I am told that only a skeleton was left by the time the battalion had passed. An order against fires did not prevent the troops from devouring steak that afternoon." Howard Hensleigh

As the First Battalion arrived at Trois Ponts the 505th Parachute Infantry Regiment was holding a two-mile sector on the left flank of the 82nd Airborne Division. The First Battalion of the 517th was to relieve the 505th. On receiving the pass down orders from the commander of the 505th, Lieutenant Colonel Boyle had his men take up 505th's positions. It was dark now the telephone wire, ammunition and mortar base plates would remain behind for the First Battalion of the 517th to use. This also would reduce any loud noises that could alert the Germans as to what was going on. The commo section of the First Battalion quickly accessed the field telephone system. Some wires were blown apart and needed to be spiced together. Bud had to crawl out into the snow on his hands and knees to find the broken commo wire. The night was dark and cold. How could anyone crawl through all of that snow to find and fix the telephone wire? Bud described it in later years; "I followed the wire until I found the broken end. Then I would tie it to something. On this night I could not find anything to tie the wire to. Then I saw a dead German lying on his back with his right arm frozen straight out. I used this arm to tie the wire to. Then I crawled further trying to find the other end. This took lots of time. Fortunately, I found the other end, brought it back to where the dead German was, spiced both ends of the wire together, and crawled back to my company's position." By this time Bud was frozen to death with only a foxhole to crawl into, but he had performed his duty. The connection was made. By midnight guides from the 505th led each 517th line company soldier to their new foxhole, dug courtesy of the 505th. The Germans never knew the two battalions had traded places. The First Battalion held this position until January 3, 1945. Orders were received at 1400 hours on the 2nd of January for an attack to take place on the morning of January 3, 1945, at 0800 hours. The other two battalions of the 517th PRCT would attack with the 551st Parachute Infantry Battalion on their right and pass through the area held by the First Battalion. The area was a thick Ardennes Forest with steep wooded hills and swampy lowlands with interspersed open fields. Trois Points was split by the Salm River, and the town was reportedly unoccupied, but the Germans held the far side of the river in strength. In the path of the attack of the 517th were the 183rd and 190th Regiments of the 62nd Volks Grenadier Division. Two hundred German soldiers had dug in at Mont de Fosse for the past five days. Reports were this German division was poorly trained, but they held the high ground with commanding views giving them tremendous fields of fire. The German soldiers were told if they could just halt this American advance they would cause the western allies to negotiate for peace. First U.S. Army had responsibility for the attack, and Trois Points would be a key point in winning the battle. The Germans were ready to put up their best defense and hope to stop the Americans. As the morning of January 3, 1945 dawned it was clear and bitterly cold. New fallen snow had drifted against fences and hedgerows offering much more cover for enemy forces. As the Second Battalion crossed through the First Battalion's position they met stiff resistance from the Germans. Company D cleared Trois Points house by house and by 0900 hours they had seized their first objective. To stop this American advance the Germans waded across the Salm River and took up positions on the west bank of the river and opened fire on the Americans. D

Company Commander, Captain Carl Starkey and a dozen others were hit with enemy fire. F Company came to their aide, and now more soldiers were dead or wounded by German fire. The Second Battalion was caught in a murderous "cul-de-sac" (half circle) of fire from the front and flanks. Lieutenant Colonel Seitz, Commander of the Second Battalion committed E Company to halt the German advance. All rifle companies were without commanders and platoon leaders. They had all been wounded and their units were left without officer leadership. This meant the non-commissioned officer (NCO) corps would step up and take charge which they did gallantly. Captain Starkey refused to leave his men until each wounded man had been evacuated. His behavior was typical of the men of the 517th.

First Battalion was now needed badly, and LTC Boyle was told to get his troops ready for an attack toward the town of St Jacques. It was late afternoon of January 3, 1945, and the long anticipated Christmas turkey dinner that never showed up on Christmas had now arrived. No time to sit down and enjoy the turkey dinner. Every man had to eat it on the run. There just wasn't time enough to enjoy it.

Bud remembered receiving a "care package" from his mother at this time. It had cookies and all kinds of candy in it. Bud didn't know what he was going to do with all of the food his mother had sent him. In just a few hours they would be attacking the Germans. Bud told the story, "I got my buddy Joe Sumptner, and told him Joe; we have got to eat this stuff quickly. I don't want any German getting the goodies my mom has sent me." So both men gobbled down the treats and prepared for combat. Bud continued, "After eating all of that food we received orders to cross an open field. We started running across the open snowy field when the Germans started firing mortars at us. Men around me were being killed on my right and my left. I learned a new appreciation for our medics that day. As I continued to run across the field, these men stopped in that snowy field while mortar rounds dropped around us to render care for the wounded. I had never seen such bravery, and always will have the highest respect for those medics. I was never so glad to be an infantryman who's job it was to run as fast as I could across the open field to safety. Fortunately I made it to our objective. I was still alive, but Joe and I had eaten all of those treats my mother had sent me. Now I had nothing to eat."

As darkness fell the First Battalion attacked toward St Jacques and Bergeval. Shadowy buildings and figures appeared. Artillery fired illumination rounds and lit up the sky. Enemy soldiers that were forming in the streets were cut down by B company machine gunners. In half an hour the town of St Jacques was secure and it was on to Bergeval. After an artillery concentration, the men of the First Battalion waded out into the snow. Enemy automatic weapons opened up. The men moved to the far edge of the village and held up there until more artillery could be sent in. Then they attacked from the flanks. One hundred and twenty one German soldiers were taken prisoner including two officers. By the time the night was over they had captured 150 and killed 40 German soldiers. The attack continued toward Mont de Fosse. German resistance began to fall apart, but the Germans were far from finished. In addition to the counterattack at Trois Points where the Germans crossed the Salm River, the 62nd German Division was now reinforced by the 18th Volks Grenadiers. On the night of January 3, 1945 the Germans deployed the 5th Parachute Regiment with about 200 men, and crossed the Salm River to take up defensive positions east of Bergeval. Colonel Graves at the 517th Regimental Headquarters recognized something had to be done quickly to stop this offensive. He ordered LTC Boyle to move his battalion to higher ground overlooking the town of Bergeval. LTC Boyle had lost many soldiers in the past few days, and knew it would not be smart to move his men in daylight. He would not take a chance of loosing any more men to German fire. Bud had experienced this first hand when just a few days earlier he and others had to run across that open field in the daytime. As nightfall came LTC Boyle moved his battalion across a large stretch of open fields that were between Bergeval and the ridgeline at 1900 hours. The First Battalion was now in place on the ridge, LTC Boyle decided to return to Bergeval to coordinate with the commander of 551st PIR. In speaking with now Colonel Boyle, USA Retired, at the 517th reunion in Washington, D.C. on Saturday, June 30, 2007, Colonel Boyle said, "I took my Intel NCO (intelligence non-commissioned officer) Sergeant Bob Steele and two other radiomen with me. There were four of us proceeding back to Bergeval. As we traversed down the ridge at night in deep snow we finally reached flat ground. Within minutes a German patrol a few feet away ordered us to stop and then gave the challenge word in German. We quickly dove to the ground. A burst of automatic fire ripped into the ground from a few feet away. I (LTC Boyle) was hit three times. Twice in my right arm and once in the left arm which was the worst because the bullet hit my femoral artery. I was bleeding badly." Just then they heard more enemy fire up on the ridgeline. Sergeant Steele leaned over to see his wounded battalion commander lying in the snow bleeding to death. He bent down and listened to the commander whisper to him, "Go to Bergeval and tell Major Fraser that C Company is in a big fight on the ridge." Sergeant Steele said, "I am not leaving you." The Colonel said, "I gave you an order. Leave me and go on to Bergeval." Sergeant Steele didn't want to leave his commander. He realized this large tall man was too heavy to carry in deep snow. He then decided the only way to get this guy to move was to get him angry. Sergeant Steele said to his boss, "You know what your trouble is Colonel, you don't have the guts to help yourself." This enraged Colonel Boyle and he came to life. "No one talks to me like that" he said. Colonel Boyle told me at the reunion, "I sent the 2 radiomen back to tell Major Fraser he was now in charge, and I struggled to my feet leaning on Sergeant Steele we hobbled into Bergeval where I was seen at the aid station by Dr. Samos. I remember Dr. Samos telling Dr. Sullivan, 'I can't get an I.V. into his arm to give him blood.' Dr. Sullivan said, 'Send him back to the evacuation hospital there is nothing we can do for him here.' Colonel Boyle said, "I was shipped out to the hospital, then to England and back to the States." In speaking with Major Fraser at the Washington, D.C. reunion on Saturday, he stated, "When I heard LTC Boyle was wounded and not coming back I quickly went to C Company and pulled them out of the battle. That next morning LTC Zais brought out Bob McMahon to be the new 1st Battalion Commander."

Sergeant Steele had just saved Colonel Boyle's life. For his bravery he received the Silver Star, and a battlefield commission as a second lieutenant (Bob Steele and Bud were friends, both attended Wilson High School in Long Beach, but the two never met until joining the 517th) (see Bud's letter dated June 22, 1945).

The fight with C Company on the ridge continued all night long. It was a tough battle. Just as dawn occurred the morning fog began to burn off. There in the fields were two destroyed German 76mm guns and a wrecked half-tracks with German dead scattered all over an acre of ground. Thirteen Germans emerged from a bunker to surrender, but the rest had escaped. Out of this German division their strength was reduced to a battalion size of about 200 men. They left fifty dead and twenty became prisoners. After this battle only seven men were left alive in third platoon of C Company.

Now with all three battalions on the high ground west of the Salm River the front lines were relatively quiet for the next two days until January 6, 1945. Snow continued to fall and men did their best to stay warm. Many men suffered from severe exposure and frostbite. Over 100 men became casualties and were evacuated because of the cold weather. LTC Boyle wounds were so severe he was now out of action, and sent to the rear. He never returned to the 517th. Colonel Graves, 517th Regimental Commander told Major Bob McMahon, Executive Officer in Third Battalion, he was the new First Battalion Commander, and to get down there and be ready for new orders to attack.

The First Battalion rearmed and regrouped until January 11th, 1945, when General Ridgway received orders to take St Vith. The 517th was attached to the 106th Infantry Division for the attack. Brigadier General Bruce Clark was newly assigned as the division commander. Little did Bud know that in thirty-seven years, in 1982, his oldest son would be an Army officer attending the Armor Officer Advance Course in Fort Knox, Kentucky. Bud's son with other officers was invited to have lunch with then General (Four Star) Bruce Clark. L. Vaughn remembered General Clark recounting how he received command of the 106th Infantry Division. He explained it was a National Guard Division and was not doing well in combat. On December 15 and 16, 1944, the 106th was hit so hard by the German attack all three regiments were destroyed. Only it's artillery, engineers, and logistical services were still intact. The National Guard Major General could not command. He didn't know what exactly to do. In fact he didn't even know how to employ radios in the division and only used field phones for communications. General Clark explained when he arrived to take over the division all radios were still in storage because no one knew how to operate them. General Clark said, "It was difficult for me to take over as the division commander as a Brigadier General (one star), when the man I was relieving was a Major General (two star)." To restore morale the 106th would be used in an attack of St Vith, and would be charged with relieving the 112th Infantry at Stavelot along the northern banks of the Ambleve.

U.S. troops fighting in the Huertgen Forest 1944, in the worst possible conditions without proper cold weather clothing during the German Winter Offensive.

Letter to Dad from Harland "Bud" Curtis

Belgium
Tuesday January 10, 1945

Dear Dad,

I'm sorry I haven't mailed this letter yet, but I no sooner had finished it than the order came down to hit the road again and since then until now we haven't stopped. It seems like years since I wrote that letter so many things have happened. I won't start to tell you anything as this past week has just been the "survival of the fittest" and so once again I've made it through and right now I am in a place where it is warm and safe and I am eating good. We even have our radio set up again and have an electric light working off batteries. I don't know how long I'll be here, but I'll write every chance I get. Don't worry about me if it goes quite a while that I don't write. It's just that I'm somewhere that it is impossible to write.

Write whenever you can too, because it is sure swell to get your letters. You would be surprised to see some of the places I'm at when your letters get to me.

Well I'll sign off for now and write mom and Jill. I guess they are beginning to wonder whats happened to me too so until the next letter.

So long for now.

Love

Bud

On the night of the 11th of January, the First Battalion moved along the line running from Spineuz, north of Grand Halleux, to Poteaux, eight miles south of Malmedy. The terrain was similar to that which the First Battalion had fought on ten days earlier at Trois Points. Colonel Graves, 517th PRCT Commander was told by Brigadier General (BG) Clark to attack at 0800 the next morning. Colonel Graves knew the danger of his men being killed in broad daylight. He decided to send his men across the Salm River that night. By morning they were already in position and surprised the Germans. It was a smooth, swift, professional performance, one of the best in the Combat Team's history. The attack continued until they seized the town of Lusnie. During the night the 106th Infantry Division, BG Clark committed the 517th to take Henumont, Coulee, and Logbierme. He must have believed these paratroopers were invincible and could do anything. Colonel Graves realized if he were to take three towns he would need First Battalion who was in reserve. Early on the morning of January 13, 1945 the First Battalion approached Henumont from the east. Artillery preparation caused the Germans to retreat and the battalion took Henumont. The First Battalion didn't stop there, but continued south passing through miles of dense woods, finally emerging into open fields with sloping hills. Company B took the town of Coulee without resistance. Company C didn't have it so easy trying to liberate the town of Logbierme. They were met with heavy German resistance and after three men were killed the company waited until nightfall to continue the attack. Major McMahon, the new First Battalion Commander arranged for tanks to support the new attack on the night of January 14, 1945, but they were not needed. Before daybreak on the 15th of January, C Company took the town without the tanks, but it cost 17 casualties for the company. The 517th continued its movement south toward St Vith. The First Battalion was assigned to set up a roadblock a mile east of "Petit Thier." There they capture German soldiers who were lost and disoriented. The regiment was now on the road from Vielsalm to St Vith. Passing through a field at night cows wounded by artillery were screaming and suffering badly. Bud could hear those cows scream and there was nothing he could do about it. These sounds like many others would haunt him all of his life. The 517th had now reached the limit of their advance and now waited for the 75th Infantry Division to pass across their front.

The 517th waited three days for the 75th to show up. This was a rag tag outfit. The unit had been formed late in the war and the Army was reaching to the bottom of the barrel to make up this division. Its ranks came from disgruntle soldiers who were part of specialized training programs such as washed out paratroopers and aviation cadets. These men were expecting to be part of an occupation army and not a combat army. They now found themselves in the most important battle of the war. The U.S. Army did everything to try and make this division successful. Each man kept bed rolls, wore overshoes, and overcoats unlike the 517th who had nothing. While waiting for the 75th to show up, Bud cleared the snow off of the ground and dug a very deep foxhole so he could keep warm. He knew he would be there for a while. He hoped that by digging this foxhole deep he would get below the cold weather at night usually 20 degrees below zero. After digging the hole Bud took tree branches, and made a cover for the foxhole hoping it would keep some of the cold air out. Ever since the winter offensive began Bud's feet were cold and wet, and after finishing this foxhole his boots were soaking wet. He decided to take his boots off that night and place them on top of his sleeping bag along with his wet socks. The next morning when he woke up his boots and socks were frozen solid. What was he going to do?

How could he go bare footed? He would not survive. Bud related this story in 2006 stating, "I had given up. This was just too much to take. I climbed out of that foxhole, and walked over to a fallen tree bare footed in the snow and sat down on the log. The weather was freezing and my feet were frozen. I sat there not knowing what to do. Then became angry saying, "If the Army wants me to go on they will have to find me some boots." Other men around him heard his despairing remarks and quickly came to his aide. Men gave him their extra socks, and so he now had three good dry pairs of socks. Bud quickly put them on, but what would he do for boots? One of the troopers found a pair of rubber boots similar to a wading boot. Other men came by and gave him some more socks. Bud put the rubber boots on and that is what he wore for the rest of the winter campaign. Bud related that as each day went on he found more socks from dead soldiers. He finally was wearing 13 pairs of socks and those rubber boots. Bud has always felt that is what saved his feet when other soldiers were being taken to the aid stations because of frostbite and trench foot. Bud described wearing these rubber boots and 13 pairs of socks like walking on pillows, but very difficult to climb trees and poles where he had to string commo wire.

With all of the advantages the 75th Division had their advance was slow and halting. On January 17, 1945, they finally arrived at the 517th location. Because of this slow advance the jaws of General Ridgway's trap closed on empty real estate. The bulk of the German XIII Corps of two divisions escaped to fight another day. The 517th was now relieved of its duties with the 106th and reassigned to the 30th Infantry Division. On January 21, 1945, the 30th I.D. along with the 517th moved to Stavelot for a period of badly needed rest and rehabilitation.

Bud finally was able to take some time to write home and did so on January 19, 1945. He explained to his mother that he was in a farmhouse just two miles behind the front lines. He keeps his letter upbeat not wanting to let his mother know of the horrors he has seen, and how terribly cold it has been for him. He does not tell her that he gave up on life, lied down in the snow to die because he was so cold and tired. Bud knew in his mind he could not take another step enduring the terrible cold weather and combat. While Bud lied there in the snow giving up to die he some how gathers the energy to get back up and continue fighting. This information had never been known to anyone until Bud was in Paris France with two of his sons, L. Vaughn and Tim. Bud was now in Paris to receive the French "Legion of Honor" medal. Now 60 years later in 2004, he could finally talk about this experience. Bud said, "I gave up and was ready to die. Then I heard these words in my mind. "You shall walk

Rubber Boots courtesy of American Patrol Company Collections, owner Raymond Meldrum

and not be weary, run and not faint." These words I remembered from scriptures I was taught in my youth by the Church of Jesus Christ of Latter Day Saints (Mormons). The words come from the book, Doctrine and Covenants, section 89 verse 20. Bud had not thought of these words since he was about twelve years of age. Why now would they come to his mind? Call it fate, call it divine intervention, call it a paratrooper's "espirit de corps" to never give up, call it what you want, but Bud got up out of that snow, kept moving, and fighting, never giving up.   It saved his life.

Rubber Boots similar to those worn by Bud when his boots froze. Bud sat down in the snow bare footed not knowing what to do next. A fellow trooper came up with Rubber Boots and 3 pairs of socks. Bud said these Rubber Boots saved his feet, and by the time the winter campaign was over Bud was wearing 13 pairs of socks. He said it was like walking on pillows.

Letter to Mom from Harland "Bud" Curtis

Front Lines in Belgium
January 19, 1945

Dear Mom,

This is about the 6th or 7th letter I have written to you this month, but we always move out before I can get them censored and on the way and then when we stop for a little while I decide to write a new letter because everything is old news in what I've written in the other letter so I tear it up and try again and then it's the same old story. I can't get it censored and in the mail. Well here is another letter now and I sure hope this one gets to you for I know you must be awfully worried about me by now. I wrote dad a couple of real nice letters too, and have been carrying them around for so long that they are wet and falling apart. I'll try and get another letter on the way to him, but if I don't I guess this one will have to do for both.

Right now I am in a small house about 2 miles behind where the fighting is going on and I'm sure glad to have this nice warm room today because there is a regular blizzard outside. It seems that you should know how cold it is here in Belgium and all the snow covering up everything because in every letter I've told you that, but then I stop to think I never did get those letters mailed so I guess I'll have to think back and write again the things of interest. Oh yeah before I forget I'll tell you what happened New Years day (again).

Well there I was slogging down a snow covered road on my way up front and dog gone, right out of nowhere there is Dean Hidreth. Boy was I ever surprised to see him and he was pretty surprised himself. He was staying in a house we passed by and he asked some guys if they knew me when he found out that it was the 517th, and it just happened that I had only passed by him a couple of minutes ago so I was only a little ways down the road and he caught up with me and walked on down the road with me for a couple of miles because I couldn't stop. It was swell seeing and talking to him again even if it was just for that short time. He was looking plenty healthy enough and has been traveling around quite a bit and now he has transferred into the same outfit that Willard Hill is in. Dean had a turkey dinner waiting for him back at the house he was staying in, so I told him he had better head on back, because if he stayed with me he would be walking all night. So I told him so long and to take care of himself, and maybe before long we could get together again for a while longer. The last I saw of him he was going back on the road to that turkey dinner. I sure wish I could have gone with him.

I guess if you have been following the news about what has been going on here in Belgium there isn't much I can tell you. I am right up where all the trouble is and it is plenty rough in this kind of weather. I guess I can't complain though. I've been up here for about a month now and the roughest fighting I've ever seen and I'm still here to tell about it so that's the main thing I'm concerned about. I've sure been lucky alright. I've seen about 140 days of combat now in 3 different countries and I can be pretty thankful for that much; even though it does get quite rough at times.

I think I've gotten most of your letters and it is sure swell to get them so keep on writing even if you don't hear from me. Don't worry about me.

I got a couple more packages and some more magazines you've sent. Thanks lots, you sure have been swell about sending me things and thanks billions for taking care of getting Jill presents for me this Christmas. I sure hope she liked that charm bracelet and things. I'll sure be glad to get some mail telling me what you did on Christmas and New Years. I sure do hope Bert was there for both days and you all had a real swell time together. Maybe next year I'll be there to. It's something to hope for anyway and maybe it really will happen.

I got a Christmas card from you yesterday dated the 23rd of December and I'm sure glad to hear Dad has gotten that journeyman job and you talked as if Bert was still there so I guess everything is going along alright. I'm sure glad.

Well I'll write again whenever I can and I sure hope this letter gets to you so you'll know that I'm okay.

Tell everyone hello and lots of love

Bud

P.S. *How about sending me a whole box of those cookies. They're sure good.*

Page 2 of a Letter addressed to Harland L. Curtis' father, written on a Sunday

Location unknown perhaps the Front Lines in Belgium

Envelope Postmarked January 26, 1945

Page 2,

Also how about sending some more of those swell cookies. I really go for them.

I have not done a darn thing today, absolutely nothing except sleep, sleep, and sleep. I don't know if I'll be able to sleep at all tonight, but I'll sure give' er a try anyway.

Some of the L.D.S. (Latter Day Saints, commonly called Mormons) boys got together today and had church of their own. I didn't find out about it until after (the services were over) *or I would have sure gone. There aren't a heck of a lot of guys, but enough to make it interesting. I just found out today that my Captain is from Utah and also a Mormon, and also my first sergeant. I feel ashamed that I didn't try myself to get a few guys together before this, but I'm glad someone did and I'll sure be there next Sunday.*

Sometime in January 1945, during the Battle of the Bulge, Bud and his close friend Joe Sumptner had the opportunity to get out of the cold weather for a while, and were in a warm farmhouse. Bud quickly took up a position on the floor, and for the first time in a long time he was able to get warm by a big roaring fire. The last thing Bud wanted now was to have to go outside in the cold where it was twenty degrees below zero. Bud began to settle down for a nice warm night's sleep when another paratrooper by the name of John Marsinko burst into the room. For some reason Bud asked how the weather was out there? Marsinko took exception to this comment, and attacked Bud by jumping on him, and yelling. Bud and John wrestled around, and fought for a while until the sergeant yelled at them to stop. The Sergeant told Bud and his friend Joe Sumptner to get out of the house and go up to the front lines. Their new orders were to relieve the men operating the communications post. Bud asked the Sergeant how he would find the communications post in the dark? The Sergeant told him to follow the commo wire on the ground. That would lead the men to the right place. So Bud and Joe began trudging through the snow on a half moon lit night, their sleeping bags dangled from the cartridge belts. It was terribly cold and very miserable to be outside. Bud and Joe walked for quite sometime and still had not arrived at the front line. Bud notices a patrol of about twenty men. He started to call out to them to ask how much further is it to the front lines? When all of a sudden something within him said, "Don't talk." Just then Bud and Joe recognized these men to be German soldiers. Twenty of them! Bud is certain the German soldiers saw them, but were too much in a hurry to get to their new location to stop and fight. Fortunately, these Germans soldiers decided to pass Bud and Joe up. Bud feels to this day the German soldiers were just as cold as they were and did not want to take the time to engage them. Perhaps the German soldiers thought there were more Americans and didn't want to take a chance of a firefight. If they only knew there were just two American paratroopers, the Germans would for sure have killed these two men. Bud knows if he had shouted out, the Germans would have killed them.

Bud's next letters to his mother comes on January 25th, and 27th, 1944. He explained to his Mom that he was put in for a Bronze Star medal for heroism while he fought in France. Then he finds out he is again put in for a second Bronze Star medal for his heroic actions in Belgium. The Bronze Star medal was established by President Roosevelt in 1944, and was a new combat award never before given prior to World War II. Many commanders did not understand its importance, and therefore many men did not get the appropriate award for heroic actions. Neither of the Bronze Star medals was approved for whatever reasons.

After the final battle was fought and the guns were silent all soldiers wanted to do was go home. To make matters worse the newly activated 13th Airborne Division arrived in Europe as the fighting stopped. The 517th was assigned to this spit and polish unit at the wars end. The 13th Airborne demanded written documentation of each unit's combat experiences. The 517th was in direct combat for 94 continuous days. The mountains of paper work that needed to be generated did not leave time for officers to write recommendations for medals. Bud was only a nineteen years old Private. How could he ask his Lieutenant if he was putting him in for the Bronze Star Medal? Bud explained in the letter to his mother all he wanted was to come home alive. His heroism would go unnoticed, as it would with thousands of other soldiers.

In his letter of January 27th, Bud, described how cold it was in Belgium, but he still felt Tennessee maneuvers were worse because he was wet the entire time. The medals were not important to him. He had enough of fighting and just wanted to come home. Because of this great discrepancy the Department of the Army recognized that many men did not receive this important award, the Bronze Star Medal. After the war the Department of Army sent messages to all units stating that soldiers who received the "Combat Infantry Badge" in World War II, could receive the Bronze Star Medal. Apparently, this message was sent out in late 1945, but there did not seem to be much interest from the veterans. They were now discharged from the Army and starting new lives. Bud like many other soldiers did his best to put the war in the back of his mind. Medals were not important now. They just wanted to forget the horrors they endured.

Now forty-seven years later in 1982 the Army reissued their Bronze Star message concerning World War II soldiers. Not until then did our family know that Dad could receive his long awaited Bronze Star Medal. At the time his oldest son was serving on active duty as an officer in the U.S. Army. He found the Army's message, and completed the necessary forms on behalf of his father. Within six months Bud received via the U.S. Mail his Bronze Star Medal, along with the Purple Heart (two oak leaf clusters), Good Conduct, European Campaign (with arrowhead and a silver campaign star signifying five campaigns), American Defense, and World War II Victory medals. It was long overdue, but Bud finally got his well deserved Bronze Star Medal. In reality he should had received two Bronze Star Medals with combat "V" for valor.

Letter to Mom from Harland "Bud" Curtis

Belgium, Wednesday January 25, 1945 (I think)

Dear Mom,

It's been so long since I've written a letter or heard from you that I've just about forgotten how to write one. I did get a couple of magazines you sent today. Thanks a lot, they sure help out.

It is still terrible cold here and more snow all the time. I doubt if my feet will ever feel warm again.

Got one letter from Jill the other day that was dated January 1st and had a picture of Dad and you and Jill dressed in those levis and plaid shirts. I don't know how that letter snuck through to me it's the first one I've had for a long time. I know you are all writing and I must have lots of mail somewhere, but it seems the mail system is all messed up as usual. I guess Christmas cards will keep coming for months. I imagine you are quite worried about me because until the other day I haven't mailed you a single letter since December. I'm awful sorry but it's the same old story of not being able to write because we were up front fighting. I hope you get that letter telling you I saw Dean Hildreth New Years Day.

Remember when I was at that tunnel in France and wrote you we were having all kinds of trouble keeping our telephone lines in with all the shells knocking them out. Not a day went by that us guys about 8 of us didn't have some close call keeping in those lines, and I just found out that we were all put in for the "Bronze Star" for our good work and then just a few days ago I heard we were put in for the Bronze Star again for the work we've done here in Belgium. Of course maybe neither one of them might not go through, but it is still nice to know that your work is being appreciated.

It doesn't really matter whether I get it or not. I'm not looking for medals or anything. All I care about is getting back alive. Boy, if those Russians keep up the way they are doing now maybe this war isn't going to last too much longer. It will sure be good to get home again.

Thanks for the things you have sent me and keep them coming. Send some more cookies and anything at all to eat.

Not much more to say. Don't worry about me. Tell everyone hello and,
Lots of love
Bud

Letter to Mom from Harland "Bud" Curtis he typed on a typewriter

Belgium
January 27, 1945

Dear Mom,

It is pretty late at night but I wanted to get a letter off to you and thank you for the three packages I got from you today. I got one with the sweater and wool hood and that package that had all the candy and peanuts and things. Boy, I was sure glad to get this sweater and hood as it is really cold and it is just what I have been needing. Where in the world did you get all that candy? It was sure swell of you to send it all to me and you really made me very happy today with all those things. Just like Christmas. It is true that the packages take some time to get here but it is sure nice when they get here so I am going to keep right on requesting things. How about sending some more of those swell cookies of yours? I really do go for them in a big way. Nobody can beat those cookies you make. I really do like them. Also, how about some more pineapple and some more honey or anything that you think to send. I guess I will really be keeping you busy but dog gone those things sure taste good and if you want to go to the trouble of sending them to me I'll sure keep on making requests for things. I even think I'll write to Jill to send me something too. Before now I didn't want to impose on anybody, but when those packages do get here and I begin munching away on all those good things it makes me want more. When we came off the front lines in France and were in a rest area at La Colle, a little ways on the other side of Nice, I met some people and they invited me to dinner, and that day one of your packages came to me and you had some pineapple in it, so I took it along that evening when I went to their place, and we had it for desert, and they said it was the first time they ever had pineapple in five years, and they told me to be sure and write home to you and tell you that it really tasted good. I had forgotten all about it until now. I really did like the people in France and along the French Rivera the climate is just like in Southern California. I would have liked to have stayed there for the duration. These people in Belgium are friendly but nothing like the French people. I guess the closer I get to Germany the less friendly they are going to be.

The mail situation is still very bad but maybe soon it will get straightened out, anyway that's what they keep telling us. I would sure like to get some of those letters that you have been writing this past month. They are knocking around somewhere in the mail or local A.P.O. (Army Post Office). One of these days I guess they will all come in at once like they usually do. I would sure like to know more about what is going on there at the home front. I feel lucky that I got that one V-mail from you and those three letters from Jill last week. I at least found out that Bert was home for Christmas and New Years and I sure do hope that he is still there right now.

For the last three nights I have been sweating out trying to see the movie show "Kismet" and every night after sitting for a couple of hours they give up hope of getting the generator going and call it off and say for sure that they will have it fixed the next night. Well I am hard to discourage and especially after this long, so I'll be right there every night hoping that it will work okay. I just hope that they get it going before we move out of this area. It has been pretty nice here this last week. No ducking any bullets or shells or sleeping in snow lined foxholes out in the woods. Boy I stayed awake many nights this last month just wiggling my toes to keep them from freezing off. Sometimes I would be tempted to drop off to sleep and let my feet freeze, but somehow I just can't punish myself that way as long as I am still able to do something about it. I've seen some pretty miserable days since I have been in this Army, but they are still none that can beat the ones I went through on those Tennessee maneuvers. That will be a year ago this February. I don't believe there was a day I was in Tennessee that I wasn't soaking wet. This weather here is just right down cold, but that is better than being wet and cold the way it was there in Tennessee. Someday all this will just be a memory like that is now though so if I can just keep on holding out and my luck stays with me I'll get back home again and forget all about this after a few days of your cooking and having that room of mine to sleep in again. I'll probably be sleeping out on the wash porch though, that is if Jill will come out and stay at the place like she did this last time. That was a lot of fun having her there, and as for that was porch it would seem like heaven itself to be there now. It would sure be great to dive into the refrigerator again. I'm going to wear that thing out going into that refrigerator again. I am going to wear that thing out going in and out of it for milk and all the other delicious things you keep in there.

I would like to get me some kind of a car when I get back if it was just to have it there when I was home, but it really wouldn't be the wise thing to do now. Might as well wait till the ration is lifted off everything and when they start making cars again so I can get a really good one. Oh well that is a long time off yet so I am not going to worry about things like that now. My main problem is getting home again and then other things can take care of themselves. It looks like these Krauts are going to keep on fighting until their whole country is overrun by Russians or Americans, and then when they don't have any place else to go *they'll have to quit. I just hope that won't be too much longer from now. Things are looking good the way the Russians are going at it, but I've learned in my time never to get optimistic about anything. I just go along expecting the worst and hoping for the best, and that way I can't be disappointed. I can't see myself getting out of this Army in two years at the very soonest. There is always the Pacific after this job is done over here. That doesn't please me in the least, but at least I'll get home again first if I make it through here okay. So I ain't worrying about that Pacific until the time comes anyway.*

I am sitting here by the switchboard again and the darn fire has gone out and I am too lazy to run down to the basement to get some wood. It is much easier just to hit the sack for the night. It is one o'clock in the morning now, anyway and time I should be asleep. I spent the best part of this evening chasing back and forth to that darn movie house on false promises. It is only around the corner though. There always seems to be something to keep me up late at nights and during the day there always are a lot of little things I have to keep me busy. I've got about thirty of those Press Telegrams (the local newspaper in Long Beach, CA) *to read. These civilians were causing us all kinds of trouble today and I was just about ready to declare another war of my own with the local electric company. They have been going around putting up new telephone poles that were downed by bombs and shells when the war was going on around here.*

I guess I had better call it a night now and get some shut eye so I'll quietly slip off now, but before I leave I want to thank you again for all the swell things you sent me. Keep them coming; they are well appreciated. Tell everyone hello for me, and I'll write again real soon.   Lots of love

Bud

Battle of the Bulge

Members of an airborne division move to reinforce the defenders of St Vith

This picture is of B Battery, 460th Field Artillery, 517th PRCT according to Richard Wheeler, B Battery, and Gabriel Delesio, C Battery of the 460th Field Artillery, 517th PRCT. At the 517th Reunion, Saint Louis, Missouri, June 27, 2008, they stated: "We recognize this picture and know it to be of B Battery, 460th, 517th PRCT moving toward St. Vith, Belgium. The picture appeared in the newspapers back in the United States." The trooper looking back from the left side of the last vehicle is PFC Odas Sweet, H Company, 3rd Battalion, 517th PRCT as identified by his son Rick Sweet.

The Battle of the Bulge ends with the capture of St. Vith, Belgium on January 27, 1945. All battalions are now ordered to rejoin the 517th Regiment located in Stavelot. Bud and others learn of the Malmedy massacre where German SS troopers gunned down hundreds of U.S. Soldiers. Colonel Peiper, the SS German commander orders the massacre. Lieutenant Colonel Seitz, commander of the second battalion is informed of this massacre and along with elements of the 508th Parachute Infantry Regiment. They are given the mission to block the retreating German SS soldiers from the city of Born. Bud hears of the massacre and wants to see what happened. Without telling anyone he leaves Stavelot and walks through the snow approximately five miles to Malmendy. There the dead American soldiers are placed in a barn. Bud walks into the barn and no one notices him. He walks through the barn looking at the dead American soldiers. Bud described, "All of them were still frozen with the hands over the heads." Bud still says today, "They were surrendering and the Germans murdered them." After seeing these bodies Bud leaves the barn and walks back to Stavelot. He does not tell anyone where he has been and they didn't notice he was even gone, but Bud would remember for the rest of his life what he saw in that barn.

American soldiers taken prisoner, and killed by German SS troopers during the Battle of the Bulge. For a summary of the Malmedy Massacre, go to the "History Place" web site.

> Archer (1985) reported, "More prisoners might have been taken, but the 3rd Battalion, 517th PRCT had just returned from the Malmedy area where it had seen, and heard of the results of the massacre of American prisoners. After Malmedy it was not easy for an SS man to surrender."-Paratrooper's Odyssey

The Battle of The Bulge becomes the largest battle the United States Army has ever fought. According to Archer (1985), "American losses were 80, 987 killed, wounded or missing. German losses were between 81,834 to 103,900." The 517th Parachute Regimental Combat Team had been in constant combat along the thirty-mile line of the XVIII Airborne Corps stretching from Ourthe to St Vith for 37 consecutive days. Between December 22, 1944, and January 27, 1945, not a single day passed without the entire regiment being in direct combat with the enemy. German attacks at Soy-Hotton and Manhay had been repulsed by the 517th. Twenty miles had been covered by the 517th from Trois Ponts to Grand Halleux, from Stavelot to Poteaux, and from Diedemburg to St Vith. During this time the 517th was attached to five different divisions and when the fighting was over the 517th had lost over seven hundred men being killed or wounded. This battle was fought in the worst conditions imaginable. The U.S. Army in Belgium was not prepared for a winter war, but they fought it. Bud would later say, "It was so cold that you never stopped moving." Bud remembered walking up to a soldier to remark about how cold it was. He talked to this soldier for a good five or so minutes before he realized the soldier was frozen dead.

Bud and the other soldiers faced one of the World's foremost adversaries, the German Army in the worse circumstances possible. It was not the brilliance of the American Generals that won this battle it was the courage and stamina of the America soldier. If American paratroopers were not there the battle may not have been won. Their courage and "espirit de corps" is what made the difference. Thank goodness for American ingenuity!

Now that the battle was over the 517th was attached to the 82nd Airborne Division at Stavelot. They were now in reserve and were ordered to rest and enjoy some of the amenities of life. Local bakeries and Food Service Companies made it possible for the men to have their first hot meals since they left Soissons France. The engineers built a shower point for the men with the luxury of hot water. The men were issued new clean uniforms. It was heaven on earth, but it didn't last long. Although it may have seemed the war might be over, it wasn't.

The 517th was attached with the 82nd Airborne Division and ordered to move out on January 28, 1945, and attack 10 to 15 kilometers through heavy snow crossing into Germany near the town of Honsfeld.

On January 29, 1945, the First Battalion was ordered to set up blocking positions to protect the northern flank of the 325th Glider Infantry. Bud and the others in the communications section went to work stringing telephone wire through the frozen tundra to provide communications from the battalion headquarters to companies A, B, and C. The 325th attacked the objective without the need of the 517th and so on February 3, 1945 the 517th was reassigned to the 78th Infantry Division at Simmerath Germany. This meant Bud, and the others quickly had to retrieve the telephone wire, roll it up, and get ready to move out. The 517th received orders to attack on February 6, 1945 to seize the city of Schmidt and the Schwammenauel Dam. From there the 517th was to move north to the cities of Kleinhau and Bergstein to relieve elements of the 8th Infantry Division. Then move on to seize the town of Schmidt near the Nideggen ridge. First, the 517th had to get to Kleihau. To do this they had to enter through the thick trees of the Huertgen Forest. This area had a reputation as being a man-eating monster. Since September 1944, four divisions, the 4th, 8th, 9th, and 28th had suffered 21,900 casualties trying to get through that forest. Now it was the 517th's turn. The forest consisted of tall dense fir trees rising out of rocky crevices and ravines. To the east was the Roer River in the center of the forest was the Kall River that had deep gorges flowing in to the Roer River. The Germans had prepared their strongest defense. It was the last chance to save the fatherland, and every German soldier knew he must fight to the death. The 517th was up against a determined force that would not easily surrender. The entire area was honeycombed with a maze of minefields, wire, pillboxes and bunkers. The fallen pine needles made perfect cover for mines. The forest was so thick daylight could rarely been seen from above. This cast a dark dreary gloom atmosphere, and with the high ridges made it difficult to find cover in case of attack. The 517th left Honsfeld and proceeded to Kleinhau. It took over 40 miles going roundabout routes to get there. The troopers were in the back of 2 ½ ton trucks shivering and frozen from the cold. By 0600 hours on the morning of February 5, 1945, all units arrived at Kleinhau. Regimental headquarters set up in the city of

Brandenburg. Bud and the other men of the First Battalion were told to set up defensive positions in the city to protect the regiment. Troops scattered out through the area digging foxholes or finding shelter in bombed out basements of formerly existing homes. The area presented a scene of war at its absolute worst. Shell craters, decimated trees and rubble were everywhere. As temperatures began to warm a little bit the snow changed to mud and that made matters worse.

At 1330 hours on February 5, 1945, Colonel Graves, Regimental Commander assembled his battalion commanders to give them orders for their next attack. The regiment was to attack that night leaving the city of Brandenburg moving through the city of Bergstein, continuing south crossing the Kall River to seize the high ground along the Scmidt-Nideggen road. It was only a mile and a half to the objective but it was covered with mines. Heavy German resistance met the Regiment's attack. German illumination flares arced across the sky revealing where the 517th was located. The Germans opened up with machine guns, mortars, and artillery fire. Men were being killed by "Schu" mines, "Tellermines", and "Bouncing Betty" mines. Just before daylight the attack was called off.

Back in Bergstein Colonel Graves along with the commander of the 596th Engineers, and the 517th's three battalion commanders met on February 6, 1945. A decision was made on the 7th of February for the 596th Engineer Company to clear a lane through the minefields.

Men from the Third Battalion would protect the engineers. The Second Battalion would then move through these cleared lanes and attack toward the Kall River. Bud and the First Battalion then would attack toward the city of Zerkall. The First Battalion's S-4, Major Hickman was directed to have shape charges and flamethrowers ready for use against the enemy located in bunkers. On an early morning on February 7, 1945, the 596th Engineers cleared a lane through the minefields so all three battalions could get through the forest. After the war it was determined that this minefield was the larges every encountered by allied forces in World War II. As the 596th cleared a lane through the minefields, German soldiers were shooting from the hilltops picking off 596th troopers one by one. More soldiers were needed to provide cover for these engineers. At 2100 hours on February 7, 1945 the First and Second Battalions prepared to go into the attack. Morale was very low. This Parachute Regimental Combat Team had been given a mission that only a full division should have attempted. Battalion strength was now that of a company. Never before had the 517th ever had to fight with such low strength. The Second Battalion led off at 2145 hours (9:45 pm) and moved through the lane cleared of mines, with the First Battalion following southeast toward hill 400. The First and Second Battalions encountered heavy German resistance with machine gun and mortar fire. The battalions were forced back. The First Battalion reassembled at hill 400, reported the situation, and waited for further orders. The battalion was one third of its starting strength of approximately 800 men. This meant Bud was with only 267 of the original men he knew in the battalion. He was lucky to be alive. Later when he was safe Bud takes just a quick minute to write perhaps his last letter to his mother. Bud knows how desperate this situation is now.

> "E" and "F" companies were now through the mine fields. Myrle Traver of "F" Company remembered what happened next. Trooper Traver said, "We moved out crawling through the snow following the tape strung by our engineers. Machine gun fire and flares were going off all around me, I was scared to death, but we kept going. We finally cleared the mine field and started climbing up and down mountain trails. Daylight broke and all of a sudden we were being shelled by our own artillery. We began to run and run, I was carrying a B.A.R. machine gun which was very heavy. When we stopped running I found myself with our Company Commander, Captain Guichie, George Flynn, and a few others. We were all by ourselves with no other soldiers around us. All of a sudden a young German soldier appeared. We took him prisoner and figured he must be only 13 years old. He begged us not to kill him, so Captain Guichie said to bring him along with us. As night fell we all took turns guarding the prisoner. While Captain Guichie was guarding the prisoner he snuck off. It started to get light and we thought we better

get out of here, but then we thought the Germans would think we would have moved from our position and we might be found so we stayed put. Next we heard some voices and saw helmets coming through the heavy brush. They looked like American helmets and so we got excited about being reunited with our soldiers. Little did we know they were German paratroopers. Their helmets didn't have the drop side like regular German helmets. As 6 German soldiers approached we were captured. The young boy German soldier was with them showing where we were hiding. Captain Guichie was a tall large man and he scared the Germans because he was so big. They were about to shoot him when I stopped them, and told them we didn't kill the young boy German soldier. I convinced them not kill our commander. The Germans agreed and so we were all taken prisoner. We were transported to various locations finally ending up at Stalag 11B in Limberg Germany. I was a prisoner of war for 64 days." Myrle Traver's complete story can be found at www.517thPRCT.org "The Schmidt Bergstein Attack – Jan Feb 1945.

Letter to Mom from Harland "Bud" Curtis

Somewhere in Germany
Wednesday, February 7, 1945

Dear Mom,

    I don't know when I will get a chance to mail this, but after just receiving such swell letters from you and 5 from Jill, I decided I was going to write anyway even if it might be a while till I can mail it. Pardon the writing I am balancing a silk hat on my lap and using it for a brace to write on.

    I'm at last on German ground and right now I am in what's left of a house. It is just a room, but it is warm anyway. It has been raining quite a bit and all the snow has melted, so you can picture about how muddy it is now. As I get closer and now right in Germany I am noticing that our artillery and air corps hardly leaves a building standing. Everything is leveled to the ground. That makes it more rougher for us guys cause when we move into a town now you are lucky to find a basement to stay in. The way our artillery is firing as close I expect any minute for this room to collapse from the concussion. The place rocks like an earthquake every time a shell goes off. I would hate to be on the receiving end. These are awfully big guns. I don't know what's ahead of me from here on. It will be rough but right now I am okay, so enough for what I am doing now.

    The letter I got from you tonight was when you said you received the first letters from me in 6 weeks. I know you have had more letters since then. I sure hope it never goes that long again. I'm afraid my letters will be few and far between for a while though, but don't worry about me.

    I'm glad to hear that Jill, her mother and all of you got a chance to get together and had such a nice evening. Sure wish I could have been there too.

    This is the first letter where you really said anything to give me any idea at all about if Bert was going to ship out again or not. I sure hope he can stay at the job he had then and can stay home. Is he still working nights?

    How does Jill like going out checking lights with Dad? I use to enjoy doing it, but gosh, that sure was a long time ago. I didn't realize I had sent so much money home. I guess I have a couple more of those $45 allotments on the way for December and January. They paid us for December the other day and I'll send the money home when I can. Oh yes, in case I forgot to tell you, when I took out that $45 allotment I had to cancel the war bonds, so you have all of them now so don't be looking for them anymore now.

As for the money I have there I still want you to use it anytime you want to and you don't have to worry about putting it back. If you find anything else to go on Jill's bracelet, be sure and get it for her. That's another thing I've been thinking about. I wish that whenever you see something you think Jill would like, nothing in particular, just anything, I would like you to take some of that money I have there and buy it. It doesn't have to be a special day either. In fact it would make me happy if you would get her something from me every week and tell her it is from me.

I wish I could be able to send more things from here, but when I find anything that would make a good souvenir it is when I am up front on the front lines, and then I am lucky just to get along myself. I couldn't possibly carry anything around. I sure packed that silk from my parachute in France along long ways though and walking across those Alps I was about ready to throw it away a thousand times. That was really a rough hike. Anyway if it wouldn't be too much trouble for you I would really appreciate it if you would get her some thing every so often from me if it is only just a corsage of gardenias. I am pretty sure you have all the money orders I have sent home now. I was glad to hear Jill finally did get that package I sent her. That skull and cross bones thing is just a warning the Germans put up over their mines.

I got that Christmas card with Harley's picture on it. He sure is a cute kid isn't he? I can just picture how proud Bert is of him. I don't think I'll have a chance to write Jill tonight, but I wrote to her last night. Tell her I got 5 letters from her today and tell her thanks for them and also for the 3 swell pictures. She sure has been swell about writing and I really am glad when they get to me.

With all the school work she must have, she must really like me to take time out and write so often as she does. I've got to go out and check a line. I'll let you know how I make out when I get back.

Bye right now.

Bud

The fighting was tense and the situation untenable. Colonel Graves reported to the 82nd Airborne Commanding General. The general immediately relieved the 517th from further actions, and by mid afternoon committed the 508th Parachute Infantry Regiment to the battle. By 0300 hours (3:00 am) on February 8, 1945, all units of the 517th pulled back to Bergstein. The 517th would recuperate and watch as regiments of the 508th, 9th Infantry Division and 78th Infantry division capture the Schwammenauel Dam and cross the Kall River. Resistance was light with the Germans leaving the area quickly. Only later did the 517th find out their mission was to draw German forces away from the main effort being made at Schmidt. The feelings of the Allied command were that if the Germans saw paratroopers committed, there must be a large contingency coming from that location. The constant pressure from the 517th convinced the Germans the American force was too strong for them. It was very difficult for the 517th to watch other units seize the objective, and victory after so much of their blood had been shed in the battle. The 517th left the Bergstein area bitter and frustrated. Archer (1985) reported, "In three days the 517th had lost over two hundred men accounting for one fourth of its total strength. The German 6th Parachute Regiment had fought literally to the death. Only one German soldier was captured and that is only because he was knocked unconscious by a grenade."

Bud wrote his last letter in combat on February 8, 1945 to his mother, although he does not know it would be his last day of combat. He briefly lets his mother know how rough it has been, and that he will be required to go out that night to lay more telephone wire. He tells her not to worry about him. If she only knew how many paratroopers from the 517th had already been killed. How lucky for him that this would be his last day of fighting.

From here on out he would not see another German combatant. In 2006, Bud described the events of the end of the battles. "Everyone was exhausted and many German soldiers were surrendering. I remembered the men in my unit giving German prisoners cigarettes and coffee." Just then Bud saw a jeep pull up. In it was General Ridgway himself. He began yelling at the paratroopers, "Get those (blank blank) cigarettes away from those Nazis. They are the enemy. They were just trying to kill you hours before. Each paratrooper quickly grabbed the cigarettes away from the Germans.

Letter to Mom from Harland "Bud" Curtis

Germany
Thursday morning, February 8, 1945
<u>Last Day of Combat</u>

Dear Mom,

Wow, what a night . It was 12 midnight when I left and said I was going out to check that line, and I didn't get back until after 5 in the morning. It is sure a job to check telephone lines in this mud. It sinks down under the mud about a foot and to find a break in the line you got to run your hand along it all the way until you feel where the break is at. You can imagine how muddy I am. This particular line I was out on last night is up to one of our forward most positions and down the valley you could see across a strip of no mans land and to where the Krauts are. It looked good to watch all our big artillery shells exploding over there on the Kraut positions. Every so often a parachute flare would go up and light up the sky for miles. Boy our artillery has sure been pounding the hell out of Jerry (Germans). It was nice to know last night when I heard a shell come screaming overhead that it was one of our own and on its way to give Jerry trouble.

I don't see how anyone could still be alive through all of that shelling, but there were sure some Krauts left down there because there was quite a bit of a fire fight going on for a while and those hienies (hinnies) were really throwing out lead from their rat pistols. Once you hear one of those Kraut rat pistol fired, you don't ever forget it. Especially if you ever have one fired right at you. They spit out the bullets so fast one bullet is practically shoving the other one out of the barrel. Our machine guns are pretty good, but they sound like a model T Ford compared to this rat pistol.

Well mom, I have written way more than I intended to because chances are I might never get a chance to mail this letter and all I wanted to do was to let you know I got your letters and Jill's and that I am still perking along okay, so don't worry about me.

I'll write you when I can and let you know how I made out tonight. I think we will be moving up on our attack tonight and that always means I've got a bit of work ahead of me tonight and chances are it might be a bit rough, but I'll write and tell you about that later.

Lots of love to all,

Bud

These past two letters were sent in the same envelope and written in February 1945. The postmark on the envelope was April 13, 1945.

On February 9, 1945, the 517th Parachute Regimental Combat Team ended combat operations. In a letter written by Clark Archer, the 517th PRCT historian to Thomas Cross he listed the final statistics of casualties of the 517th PRCT. This information was obtained from the 517th PRCT mail call email site on February 9, 2005.

Sixty (60) years ago today(2005) February 9th was the last day in combat for the 517. The 517th PRCT had 1,500 casualties. 1,400 were in the 517th Regiment---70% of its original authorized strength.

On the next page is a letter from Clark Archer to Colonel Tom Cross explaining what it cost the 517th in lives, the roughness of the Battle of The Bulge, and the fighting on German soil.

"Colonel Thomas R. Cross

Dear Tom,

While statistics are quite boring they can be used effectively to demonstrate 517 PIR combat efforts during the Battle of the Bulge:

1. During the period of 20 Dec 44 through 10 Feb 45 the 517 PIR had net losses of 63 officers and 888 Enlisted Men. Our 9 rifle companies suffered 714 battle causalities and were reduced from an average strength of 156 to 77.
2. December and January operations near Soy, Manhay, Trois Ponts, Bergeval and Saint Vith resulted in a total of 698 battle casualties.
3. During January 45 the Regimental Aid Station moved 5 times and treated a total of 1,122 men. They treated 293 shell fragments, 104 bullet and 69 concussion wounds. Medical reports on these men can be located in records of 662nd, 331st, 105th and 307th Clearing Hospitals.
4. During February operation in Germany the percentage of men killed in relationship to total wounded rose dramatically from 12% to 20%." Clark Archer

Letter to Dad from Harland "Bud" Curtis

Northern France
Monday, February 19, 1945

Dear Dad,

I got your letter answering that V-mail I sent you when I was there in Belgium. It is rather late at night, but I want to get this letter written before something comes up. I have been leading a pretty soft life this past week and although I have had plenty of time to write letters I just haven't done it. I have been spending a lot of time reading all those Colliers magazines (a magazine of current interest similar to a People magazine of today) you sent me and in the evenings they have a picture show. I just came back from the one tonight. I have just been fooling around like that all week and haven't really accomplished a thing except getting fat on the swell chow they have been feeding us. I have eaten better this last week than I can remember for many months.

I hated to hear that Bert was going to ship out again. I sure hope it won't be for along. It would be nice if he could get a ship coming over this way for a change.

I haven't seen Dean Hildreath since that time New Years Day in Belgium. I know where he is at, but I guess I hadn't better say in this letter. I might get to see him again though.

The weather here has been perfect. It has been a little bit chilly, but it's along ways from being cold. I guess it is just an early spring. I think this must have been an extra early spring for Belgium.

It was just colder than heck and then all of a sudden before you actually knew it, all the snow had melted completely away and left things pretty muddy for a while. A couple of days ago the sun was really putting out the heat and it sure felt swell. I should have taken some pictures but I was just too un-ambitious to do anything. I must have spring fever or something I guess.

Let me know as soon as Bert does ship out and I will write to him. I've got his address. Where is Lorrain going to stay now? Will she go back to Redondo or stay there at the house. I sure like Harley's picture on that Christmas card. He really is a cute kid alright. I imagine you are all quite proud of him. I sure am.

How do you like your new job? Mom says they have been keeping you going all hours. I guess you are being kept pretty busy. Say, what does a guy have to do to be an electrician anyway. It seems like everyone in the family is and I ought to plan on some sort of job if I ever get out of this Army. I don't know much about what I really want to do, but what ever it is I want to be my own boss even if its pushing a peanut wagon around town. I've had enough of this taking other peoples orders. It seems that between the 3 of us, you, Bert and I, we ought to be able to set up an electrical contract business, but of course I don't know a heck of a lot about what steps a guy would have to take for that. It was just an idea. I imagine after the war there is going to be lots of people wanting to build and lots of electrical work to be done. Well I'm not going to worry about that. I can't see any possible way out of the Army for at least another 2 years at the soonest. That is not a very bright outlook as far as I am concerned, but my feelings are shared with several million other guys that are all in the same boat.

Well the guys are all in bed now and say the light bothers them so I'll sign off for now. I should be in the sack myself. Reveille is just a few hours off. Tell everyone hello and thanks for the very interesting letter.

Lots of love to all,

Bud

P.S.

Can you get a pin that holds the strap on this watch you sent me. I doubt if I can get one like that over here so if you can find one for this make of watch just send it in a letter. I could sure use it. I really took a beating on loosing things in that push through Belgium. I even lost this watch, but got it back again thank goodness. I was really feeling bad about loosing it. It keeps perfect time. It really is a honey I've still got this pen you sent. I had it in my duffle bag and got it out when we came back here. I lost the pencil that came with it. I was in a house one night and Jerry (Germans) counterattacked us and things were in rather a mad scramble and I had to pull out fast and left everything behind except me and my carbine. The krauts blew the house out with mortar shells. We had quite a time that night, but I'll tell you about that some other time.

Good night again and thanks for everything.

Love Bud

On February 15, 1945, the 517th was notified by Headquarters, XVIII Airborne Corps, that the 517th Parachute Regimental Combat Team would be dissolved. They would now become a parachute regiment under the 13th Airborne Division. This came as a big shock to the 517th. All through the war they had acted independently of any division. They were there own division unto themselves in a manner of speaking. Now the 517th was ordered to report to Joigny, France to join up with the 13th Airborne Division. On February 21, 1945, the Combat Team was officially dissolved. Bud and everyone else was part of the new 13th Airborne Division that had not seen any combat. At Joigny the troops moved into local casernes (barracks). Marching and training resumed slowly. New clothing was issued with orders to sew on the Blue shoulder patch with Orange Unicorn on the left shoulder of their uniforms. Bud now has time to write a nice long letter to his mother. He does not talk about any of the combat or horrors he has seen, but lets her know how nice Joigny, France is. He tells her it is the best he has ever seen since being in the Army. Bud and the other soldiers know this treatment cannot last long. A war is still being fought in Germany and when it will end no one, at least at his level knows.

Letter to Mom from Harland "Bud" Curtis

Joigny, France
February 25, 1945

Dear Mom,

It is Sunday morning and after that swell breakfast of fresh eggs and French toast I just ate I feel like going back to sleep for a couple more hours instead of writing, but I haven't written for a few days now, so I hadn't better put it off any longer. The weather here is swell and the sun has come out everyday that I've been here except yesterday of course because I was all dressed up and was planning on taking some pictures, so the sun just hid behind the clouds all day like it does so often in California when you are wishing it would come out. Gosh, but I would like to be home again on a Sunday morning and having you rush me around trying to get to church on time. That was quite a job for you I guess. I can't remember when I went to church last unless it was in Atlanta, Georgia when I looked up one of our churches and went. The Army has regular church services, and even up on the front lines they try their best to provide someplace where they can have church services but it doesn't seem the same to me so I never do go to them. I just wait until I get back to the States again and can go to our own church where I can have the right feeling of it all, but until then I have my own way of thinking and going to other churches just doesn't agree with me.

This place where I am at now is the best deal I've had since I've been overseas. We have a nice clean place to live in with bunks and all. We made the bunks ourselves and have mattress covers filled with straw, but it is like sleeping on an innerspring mattress compared to the other places I've slept. I am on the third floor; that is the only thing rough there is about this place. Running up and down those stairs is going to get the best of me yet. There is another building for showers and I have been getting my share of them alright. They have a picture show every night and there is a town just around the corner, so there is enough to keep a guy busy if he has any spare time. The chow is the best they have ever fed us yet so all in all I am being treated swell and am quite contented with everything. I saw Jack Dunaway at the show the other night. He is looking healthy enough. I don't know where Dean (Hildreath) is at now. He is with the 82nd Airborne now and I haven't heard anything about them since we left Belgium.

I got the letter Dad wrote me with Eddie's address in it (Eddie talked Bud and Garth into dropping out of high school and going to San Francisco to live in 1942). I'll have to write Ed a letter today and see how he is making out. It has been over a year since I've heard anything about him.

I got a letter from Phyllis Prouty and she is still going to school in Illinois and she said to tell you and Dad hello.

Well I think that about covers everything for right now so tell everyone hello for me and I'll write again soon.

Lots of love,
Bud

P.S. My new A.P.O. (Army Post Office) is 333

Letter to Mom from Harland "Bud" Curtis

Joigny, France
March 11, 1945

Dear Mom,

Just a thank you note to say thank you for the two swell packages I received from you today. Boy that fruit cocktail sure tasted good along with those delicious cookies. That makes

4 packages I've got from you this week. I've got quite a supply of in between chow...chow (no I'm not stuttering). I'll I have to do is reach under my bunk and "Presto" out comes a delicious cookie or candy bar just like magic. Thanks a billion Mom. I sure do appreciate the trouble you go to in sending me those swell things to eat, and I am sure you would be very happy to see how happy I am when I get your packages. I am even going to start making more requests like you asked me to. How about sending some more fruit cocktail? It sure hit the spot.

Love Bud

March 12, 1945, the 13th Airborne Division has been ordered to make a combat parachute jump into Germany. The operation will be conducted by the First Allied Airborne Army, called Operation VARSITY. Bud and the other soldiers thought just for a moment that their part in combat was over. Now one more combat jump! Would he survive? With all that he had been through how could this be happening now? Bud and the other men prepared for the combat jump. Bud and the others did not want to be under the command of the 13th Airborne Division who had never been in combat. These guys were looking to prove how tough they were and that kind of thinking could get a whole bunch of men killed. Bud and others had just gone to sleep when they were awakened and taken to the airfield. Bud and the others were in full combat gear loaded on to the C-47 airplanes. Engines were running and Bud knew it would only be a short time before he would be parachuting into Germany. Then all of a sudden the airplane engines were shutting down. What was going on? A sergeant entered the plane and said that operation VARSITY was called off. Bud and others were relieved that they did not have to make another jump. They returned to their barracks and stood down.

What would happen next nobody knew, but it was not long before another mission came down for the 517th. Sure enough the 13th Airborne Division received orders to be part of operation CHOKER. This mission was to cross the Rhine River in support of Seventh Army. Again preparations were made. Bud and the others found themselves once again loaded on C-47 airplanes in full combat gear. This time they would go. They would do the mission, they had too. No one could doubt his luck could hold out twice. But it did. Once again the airplane engines shut down and the mission was called off. General Patton's Third Army had crossed the Rhine River and on the attack well into Germany. By this time Bud's nerves were shot, like many others. How many times would they do this to him? How much could he and the others take? Bud writes his mother on March 26, 1945, and expresses how lucky he is to be in Joigny, France. He tells her he feels guilty for living so nice like rear echelon troops. This type of treatment is not normal for him and he and the others of the 517th expect to be back in combat any day. It is what had happened in the past and he does not want to get his hopes up that perhaps he just might survive this war.

Letter to Mom from Harland "Bud" Curtis

France
March 26, 1945

Dearest Mom,

Just a short note before the nurse comes in here and turns out the lights. I don't know what's the matter with me. I have all daylong to write, but doggone I just can't get in the mood to write until night time. I have always been that way. Even at Camp Mackall I would seldom ever write a letter in the day time, but then I would stay up until way late at night to write to you and to Jill.

I haven't had a single letter from anyone since I came here to this hospital almost two weeks ago, so you can see I don't know much about what is going on at home. It looks like I'll be here for another week or longer so when I do finally get back to the outfit I will have a lot of your letters to answer.

I sure wish I could have someway to get you something for Easter but I guess all I can do is send my love and hope it won't be too much longer until I can be home everyday.

The rest I've had here has really been swell. I feel like a different person altogether from getting a good night sleep every night and clean white sheets too. How about that. The chow here is swell. I don't see how I could help but to have gained weight since I've been here. I have read quite a few books. That's about all there is to do. They have a movie about three times a week. It would be nice I guess to be in a rear echelon outfit and live like this all the time, but anyway I'll never feel guilty of not having done my part in getting this war over with, and I'm plenty proud of those wings and boots you have there at home. It is not everyone that can have them.

Time to say good night.

Lots of love and kisses –

Bud

On March 30, 1945, President Roosevelt went to Warm Springs, Georgia after a long exhausting trip to Yalta where he met as the "Big Three" with Churchill, and Stalin. On the afternoon of April 12, 1945, President Roosevelt was carried to his bedroom saying, " I have a terrific headache." Doctors attending the President said he had a cerebral hemorrhage and he later died that day. Throughout the world the news was reported President Roosevelt had died. Bud and the entire world were in shock. Bud had doubts (as many others did) about Harry Truman leading the Nation.

On April 4, 1945, the 13th Airborne Division received orders to move to airfields in northeastern France. Intelligence has been obtained that diehard Nazis, SS troopers and such would carry out an attack from the Bavarian Alps down into France (SS translated is Schutzstaffel which means Protective Squad. Originally the SS provided bodyguards for Nazi Leaders). To counter this attack the 517th will parachute into Stuttgart to block the movement. Once again the mission is called off. April and early May would leave events that would change the world forever. President Roosevelt died, Hitler committed suicide in his Berlin bunker and the Germany Army surrendered to allied forces on May 8, 1945.

Chapter 14
THE WAR IN EUROPE ENDS

On May 12, 1945, the 517th returned to Joigny France for the last time. Combat for the 517th was now over. Since arriving in Italy on May 31, 1944, and preparing for their first combat mission with the 36th Infantry Division on June 18, 1944, a great deal of action had occurred for the 517th Parachute Regimental Combat Team. They had not been assigned to a Division since they left the 17th Airborne Division at Camp MacKall, North Carolina in May 1944. They were the only parachute regimental combat team in the European Theater of operations. They were assigned to any divisional size or larger unit that needed help in a combat mission. This is probably why they spent so many days in direct combat. 94 days on the front lines. No other combat unit in World War II or since can make that claim. In today's terms they would be known as a "911" force. When some unit needed help they were there to fill the gap. They were assigned to eleven different units. The 17th Airborne Division "state side." In combat, the 36th Infantry Division, 1st Airborne Task Force (combat jump into southern France), XVIII (18th) Airborne Corps, 3rd Armored Division, 30th, 75th, 106th, Infantry Divisions, 82nd Airborne Division, and the 78th Infantry Division. During these assignments the 517th fought the following German units. 1st and 2nd SS Panzer Divisions, 29th SS Panzer Grenadier Division, 5th and 6th Parachute Regiments, 116th Panzer Division, 34th, 148th, 242nd Infantry Divisions. The 18th 62nd, and 560th Volks Grenadier Divisions and finally the 162nd Turcoman Division.

Now the war was over and someone needed to record it for history's sake. The 517th found its self being deactivated as a Parachute Regimental Combat Team, and reassigned as a regular parachute regiment under the 13th Airborne Division. Bud and the others hated the 13th Airborne Division because they were a non-combatant division. The 13th Airborne Division patch was blue with a gold unicorn, and every man had to sew this patch on the left shoulder sleeve. Now the real fun was about to begin.

Howard Hensleigh remarked on "Mailcall" # 1265, February 12, 2007,

"The 517th started out as the parachute infantry regiment of the 17th Airborne Division under General Miley. We wore that patch. We were pulled out and made a Parachute Regimental Combat Team for Italy, Southern France and all of our fighting in the Bulge and Germany which ended at Bergstein. We never did have an official patch of our own although Dick Spencer's Battling Buzzard has followed us all these years. In Southern France we were the largest element in the First Airborne Task Force, but we did not wear its patch if it had one. When we reached Northern France we became 18th Airborne Corps troops under Mat Ridgeway and wore its patch. Before the war ended in Europe we were transferred to the 13th Division where we had several missions such as the Rhine crossing that were canceled because of rapid Allied advances. We wore the 13th Division patch, most of us reluctantly. The 13th Division that did not see combat resented us almost as much as we resented them. We went with the 13th to the States on our way to Japan. The trip to Japan was cut short while we were in the middle of the Atlantic by the Japanese surrender which ended the War. Before we left France, at Joigny (south of Paris), high point men were given the opportunity to leave the outfit and go to other units. Some went to the 82nd, and some I'm sure to the 101st."

The 13th Airborne Division demanded administrative perfection. Morning, status, and strength reports had to be perfect. Every "i" dotted, every "t" crossed. Haircuts, uniforms with ties, spit shinned boots, calisthenics, and yes marching, and plenty of it. This was busy work to keep the men active and physically fit. The 517th was a combat unit, not a paper pushing garrison unit. To them this was harassment.

Because of the enormous amount of paperwork men who could type were requested to volunteer for administrative duty. They were needed to complete the required documentation to record the actions of the 517th. Bud had observed the clerk typist in his company, and determined that he had a pretty good life. Bud knew how to type and decided to volunteer for the duty. He was promised the position of acting sergeant with a promotion to corporal as soon as he knew the job. Bud quickly got with the corporal who was the company clerk in First Battalion, and started typing. Soon thereafter the corporal left and Bud was left on his own to perform the company clerk duties. Bud felt this job would keep him alive, and although the war in Europe was over there was still a war going on in the pacific with Japan. Bud felt he might have a better chance of staying alive fighting the Japanese if he was the company clerk. Bud and the other men knew as good as the 517th was it was just a matter of time before they were ordered to fight in the pacific. While troopers took passes to Paris, Brussels, and London, Bud slaved away typing reports. Not once did he go on any of these trips, but kept his nose to the grindstone anticipating his corporal strips, and then maybe sergeant.

Soon information came to the men that they would be given a choice to stay in Europe as an occupational force, or volunteer for combat duty in the pacific. As you read Bud's letters you will see he agonizes over this decision. If he stayed in Europe he would be safe, but the garrison duty of the "spit and polish" 13th Airborne Division was taking its toll on Bud and the other men. He decided he wanted a transfer.

In reading Bud's letter dated May 29, 1945, you can see he giving his opinion about information his mother wrote him about his friend and cousin getting back home sooner than anyone else. Bud expressed his displeasure about his friend Dean Hildreath, who served in the 101st Airborne Division, and his cousin Boyd Johnson, who served in the Army Air Corps. They have gotten transferred home soon after the war ended because of "War Nerves." Bud was very upset about how they got home, and finally wrote a little about his war experiences to his mother. He expressed how much worse it was for him than what these two men went through. He like other soldiers wanted to get home from the war also, but he was not willing to use the excuse of "War Nerves". The esprit-de-corps taught him in the 517th would not allow such disloyalty.

Letter to Mom from Harland "Bud" Curtis

Joigny, France
May 29, 1945

Dear Mom,

I am back with the 517th again and I've received quite a few letters from you and Jill. I won't be able to answer them all, and no doubt I'll forget to answer some of the things you have asked me, but I'll try to get most of them. I wrote Jill a long letter last night and explained most everything of where I have been and all that, so I imagine she will tell you and anyway I haven't too much time right now to write. I also told Jill last night that there is a good chance of us coming home in July, but as I've told you before anything can happen to change those plans. Anyway it is something to think about and I sure hope things work out so I'll be home before Jill's next birthday.

I got another letter you sent me that was returned to you from the hospital. I can understand why they sent those letters back to you. It would have been much easier if they would have sent those letters back to the 517th. Some letters they did and some they sent back to you. I am pretty well mad at whatever dumb kind of a jerk that was responsible for that.

I got another package from you yesterday and one from Phyllis Prouty. Maybe it would be good idea at that, if you didn't send any more packages for a while. When I find out for sure what's cooking about if I am going to the States, Pacific or Occupation Army, I'll let you know. Right now though everyone is willing to bet we are going to the States, so hold up on the packages for a while. I'll let you know latter if I think you should start sending them again. Thanks. Phyllis says to tell you hello and stuff, she is still in Illinois.

I am afraid it is too late for me to develop "war nerves" now. Maybe I can get "inspection jitters." This isn't the E.T.O. (European Theater of Operations) anymore. It is the E.T.I. (European Theater of Inspections). To be honest with you Mom, I haven't got very much respect for either Dean Hildreath or Reid if that's why they were sent home. WAR IS HELL, I've got other names for those kinds of guys. Of course some guys really do go nuts I guess, but I know damn well Dean didn't see a tenth of the combat I have. I had over 100 combat days of fighting and being shelled when I first saw Dean on New Years day and he was way behind the lines then. He told me that was the closest he had ever gotten to any place where the fighting was going on. I don't know what he did after I saw him, but it couldn't have been too much. As for Reid, you can tell him for me that I would have taken his job as any guy in the air corps job for a whole week to each day I spent in ground fighting. It is a good thing that Dean and Reid weren't with me on that jump into southern France. Ask them how they'd like to jump out of plane miles behind enemy lines and land in the middle of a German patrol. Ask them how they'd like to be shot at just 20 feet away looking down a kraut gun. Ask them how they'd like to crawl under machine gun fire in the snow and seeing their buddies blown to bits by shells. It is a darn good thing they weren't with me when I was at that tunnel at Col de Braus, France. They would have really gone nuts. For seven weeks, day and night we sat on one mountain and the Jerries on another one and listened to screaming shells and sweating out when it was you that was going to get blown to hell instead of your buddies. Maybe I would have gone nuts myself if we'd stayed there much longer. It is those screaming shells that drives a guy nuts. Don't get any idea that I wasn't scared, because I was. Every time I would stick my head out of that tunnel to go out and fix a telephone line that was continuously being blown out, I knew my chances of getting back again in one piece were pretty slim, but I never got so scared that I would turn yellow and quit. You say it is War Nerves, but I say they are yellow. Sure it would be great to get home, but not that way. I've still got a little pride in myself. I'll never have much respect for Dean or Reid. I was pretty lucky all the way through this war. I never told you before, but I did get nicked a couple of times, but not much more than a minor cut. Several times I could have put in for a purple heart, but never did because I figured some guys were getting killed to get that purple heart and I didn't think I deserved it for just a cut. I could have got a purple heart for one time in Belgium. I was in a house and a shell exploded right outside the window I was by, and showered glass all over me, but I just got small cuts and nothing to bother about.

Well now they want all guys that got any kind of a wound or cut no matter how small to put in for the purple heart, so I did today. If it goes through it will mean 5 more points and I think the state of California pays $300 for every soldier with the purple heart so that wont be such a bad investment. I still don't think I deserve it or anyone else that got them for minor things when some guys have to get killed to get them. Another thing that makes me mad is the way officers put each other in for the Bronze and Silver Star medals when they never got with in miles of the front lines. A WAC (Woman's Army Corps) in Paris got the Bronze Star Medal. I don't know what for and my imagination cant stretch enough to think of anything anyone in Paris could get a Bronze Star Medal for let alone a damn WAC. I know an officer that got a Bronze Star Medal for setting up an efficient mail service.

How about that!! The air corps is a good place to get medals because there are so many officers and plenty of graft.

I've got a lot of beefs about that air corps not counting about the time they strafed and bombed us in Belgium and how they dropped a whole company of paratroopers in the ocean on that southern France jump. Well that's enough now for my complaints. I've got lots more, but to heck with them. No I might as well tell you what I think of this point system. The two things that make me the maddest are, number 1, they say "you men chose this point system yourselves." The heck we did. They didn't ask me anything about it and they didn't ask millions of other guys about it. In fact I am willing to bet no G.I. was asked what they thought about it. Number 2, Is this giving 12 points for children. 12 points for something that had nothing what so ever to do with winning this war. Sure I am all for guys to get back to their families, but they shouldn't give that many points for kids and the only ones that should get credit are children born before Pearl Harbor. Okay, now, number 3, is campaign stars. Guys like me fought to get those 5 points battle stars while other guys got the same stars for just being in the area of the campaign hundreds of miles behind the lines.

One incident for example is a headquarters in London that got the same stars I fought day and night for and they got it for one man signing a piece of paper putting a café off limits 900 miles away from Belgium. See what I mean? Well, I am pretty well fed up with it all and I'll guarantee you if I ever do get back in the States they will never get me out again. I will have a permanent broken leg or something. Why should I get myself killed when I can get the same credit in a non combat outfit? I could get points faster by staying in the States and raising a family. You know how they talked about giving the man that did the fighting the most consideration when the fighting ended. Well you are seeing what a stab in the back they gave us now don't you? Nope, I have made up my mind Mom, my fighting days are over even if I have to take 20 years in prison, but I wont, I can't be yellow too. I guess, and in my case my conscious won't bother me a bit. I've done my share of the fighting and didn't get any credit for it. I'll stick around here with the 517th for now because I think I've got a good chance of getting home, but if I find out they aren't going home, I'll do something to put me back in a hospital and buck for a section 8 discharge. Those hospitals are okay. If I'd known how they were going to stab me in the back and all the guys that did the fighting I would have spent all my time overseas in a hospital. I guess I sound pretty bitter, but that's just exactly how I feel. What do you think? Am I right or not?

Well enough for now. I didn't intend to say as much as I have.

So long for now.

Love Bud

Letter to Mom from Harland "Bud" Curtis

Joigny, France
May 31, 1945

Dear Mom,

No mail today but I got several from you a couple of days ago and from Jill too. The weather has cleared up and the sun was out nice today.

I have written to Jill several times the last couple of days and I don't know if she told you or not but an order came out a few days ago that all men that fought in Italy and in France will not have to fight in the Pacific. Well they can always change that order or just say the Army needs you in the Pacific and that's that. Well anyway the score seems to be shaping up to where they are going to give us a choice of either going to the Occupational Army and stay over here and maybe not get home for a year or two from now but you could be pretty sure of getting home alive and that's something to stop and think about. The other choice would be to go to the States not long from now and then get a 30 day furlough.

Then maybe a couple of months or so of training in the States and then it would be to the Pacific from there.

Well I don't know mom, that's going to be an awful decision for me to make by myself. You got to look at both choices from all angles and right now I don't know what I'm going to do if they come around and say "make a choice." What do you think Mom. I wrote Jill tonight and asked her what she thinks I ought to do. I hope I get your letter and hers before I have to decide by myself. Being over here now isn't so bad. The chow is much better since the war ended and once they get things organized in Germany all there would be to do is sit around and see that no trouble started up. In the meantime there would be chances to go to school over here, but that couldn't be too much because the teachers will just be guys from the outfit that know a little bit about some certain subject. Like the guy who will teach shorthand. He is my sergeant here in the Wire Section and the guy who is going to teach French. I know as much about it as he does.

Naturally, you are going to learn something but it might be the wrong way. These guys may know the things, but it is not like an experienced teacher who knows how to go about it the right way of teaching. Well anyway besides that the guys in the Occupational Army would get passes to Paris, Brussels and big cities at regular intervals and also a furlough to Nice on the French Riviera and to England occasionally. There is not getting around it that a guy would have to have pretty good reasons for wanting to get home to turn down staying over here now. Well I do want to get home plenty bad that's just the trouble, but after those 30 days it would mean going to the Pacific. I think I could face hell itself if I could have 30 days at home. I explained to Jill how things would be if I came home now, and she has a lot to do about me making up my mind as to what I should do. I just hope I get her answer before I have to decide for myself. Heck the way plans have been changing left and right here lately there is no telling but what something might come up to change these.

They had it pretty well settled last week that we would all be coming home in July, but now these new ideas that came out has changed all that. They might even have to break the 517th up before it's all over. There is about a third of us left that fought in Italy, and a third they joined us throughout France, Belgium, and Germany. There is about a third of the guys that joined us here when we came back from Germany, and those guys just came over from the States and had never seen any combat at all. That's how things stand. There is not many of us guys left that came over on the same boat together a year ago.

Write soon and let me know what you think I should do if I have to make this decision of staying her or going to the Pacific by way of the States

So long for now,
Lots of love Bud

Letter typed to Pop from Harland "Bud" Curtis

Joigny, France, after the war
June 22, 1945

Hi Pop,

I got your letter today with all those swell pictures. Gee whiz, but that Harley has grown. Is that really him or some neighbor kid you picked up somewhere? At this rate by the time I get home he'll be sweating out the draft board. Those really were swell pictures. That bathrobe of mine kind of drowns Irene, doesn't it? What do you think of Irene anyway? (Irene befriended Bud while he was on a weekend pass from Camp Mackall to New York City just before Bud left for the war. Bud always referred to Irene as a "big sister." He told her if she was ever out in southern California to stay at his folk's house. Irene did go to southern California and did stay at Bud's folk's house). I got a nice long letter from her the other day that she wrote while she was out to the house. She says she was having lots of fun, and went out to church and met all the relatives. What do they think of her and vice versa? I don't know whether to write or not to her because she said she'll be going back to New York pretty soon now and if I wrote to her she might already be gone and not get the letter. I guess I had better wait until I get her New York address. I'm sure glad that she came to California because at least she did get out to the house and I know you treated her swell and did everything to make her happy. I promised her if she ever got out to California that I would try to show her a good a time as she did me when I was a stranger there in New York. She really was swell to me there and took me all over the city. I guess she has told you all about it though. I guess the next time I see her again it will be in New York.

    I wrote Mom a letter last night and told her I would be coming home the first week in September. That still give me quite a bit of time left over here. Oh yes, Irene said that Jill came out to the house while she was there. What was the outcome of that? As yet I haven't heard from Jill. I guess I won't be hearing much from Jill anyway the next three weeks or so, because the last letter I got from her she said that she was going back to Utah for a while. She also said in that letter that she wanted me to stay here instead of going to the pacific. You also told me the same thing in this letter I got from you tonight, and I guess I'll be getting another letter pretty soon from Mom telling me to stay here too. Gee whiz, don't you people want me to come home?? I thought it all over and I decided that I would rather take a chance on getting home even if it does land me up in a foxhole out in China someday (When the war ended the 517th PRCT was assigned to the 13th Airborne Division. This division had never seen combat and wanted all of the men to conform to state side protocols such as shinned boots, haircuts, polished brass, marching, etc. For hardened combat paratroopers these routines were nonsense. Many men were ready to do anything to get out of the 13th Airborne Division. Bud volunteered to go to the pacific and fight Japanese). Heck, I might as well see how the people live on that side of the world. The way things stand now though maybe I'll get to stay in the States longer than I think and maybe not. When we get to the States they are going to hold us in strategic reserve, and that can mean anything. It could mean we would never leave the States unless something real serious happened in the pacific so that they would need us. It just mostly depends on how things in the pacific develops I guess. Well I like it lots better to know that at least now they are planning on trying to keep us in the States as long as they can. My purple heart and a 5th campaign star for the battle of central Europe came through today and that gives me "ten" more points. I've got 67 points now and if I can do it I'm going to try and get a cluster for that purple heart. You see there were lots of times I have been in houses and had a German shell come in and practically blow the house out from under me. I always got out okay with maybe a slight cut from a piece of glass or banged up a bit by rocks and stuff flying around. At the time that didn't mean anything to me because I've got more bruised and cut up in a good game of football or just fooling around, but not everybody is putting in for the purple heart for all those little cuts and raising there points score up 5 points a whack. I don't like putting in for a purple heart for some little minor cut, but that's what all the guys are doing now, and some of the guys are putting in for the purple heart that never even got anywhere close to up front, so I figure I've got better reasons for getting a purple heart that half of the guys that are getting them. All these medals and decorations are just getting to be a big joke now. It really makes me mad, because there are some guys that really earned these decorations the hard way. The only thing I have any respect for now is the little bronze star that I've got in the middle of my wings. That star can only be worn by guys who actually made a combat jump into enemy territory. Any paratrooper you see that is wearing a pair of wings and has a little bronze star in the middle of them you will know that he has made a combat jump. That's the only thing I've got any respect for because the guy actually had to make a combat jump to get it.

    These other stars that you see guys wearing on their service ribbons are mostly just a big joke. They used to call them battle stars, but they had to change it to "service star" because most of the guys wearing them have never seen a battle in their life. They do that so that guys who sit behind the lines can get points too. Just like the ground crew in the air corps. Those guys will never see any fighting, but the planes they serviced did so therefore they get this star. I fought hard for 4 of the stars that I am wearing, but that 5th star is just a joke. When that jump was made over the Rhine river last March 24, 1945, the 517th was sitting at the airport ready to take off at a minutes notice at anything they were needed, but the way things turned out the 517th was never needed, but yet we got the same star for being in reserve way behind the lines while those guys that did jump were out there earning it the hard way. See what I mean? Well that's the way it was while I was out earning those other 4 stars, the hard way. Some other troops were sitting way behind the lines and getting the same credit. It is just a big laugh. None of these things mean anything to me except the 5 points. Oh yeah, the French presented us with some kind of decoration. I forget just what it was, but it entitles us to wear a braid of some color that goes around the left shoulder I think. Heck I'll be lit up like a Christmas tree. So far I've got my wings with that star in the middle, the ETO (European Theater of Operations) ribbon with 5 stars on it, the good conduct ribbon, the purple heart and the combat infantry badge. Oh yes, and that French Braid that we'll be getting pretty soon. By all rights I was suppose to get a Bronze Star medal for action above and beyond the call of duty and all that sort of stuff, but somehow or another someone snaf-fooed it I guess. Probably too many officers in for it at the time. Well that is the way it goes. I don't give a darn for the Medal, but I would like those 5 more points. The officer that put me in for it is in the hospital, so there is not much I can do about it now. I guess I can chalk up some more ribbons and stuff when I get to the pacific. Heck, I don't know where I would put anymore junk on me than I've got now.

  I sent the big box home last week. All I've got in it is junk so don't think it is a French Ford V-8 when you get the box. I've got some of Jill's letters in it, and the sweater and things Mom sent me to wear when I was in Belgium, and also I put in a couple of German officer field packs that I picked up in Southern France. They got fur on them and might look good on Bert's motorcycle (See packs on page 187). That is what I sent them home for. See if they'll fit or not, and how it looks on there....

  It is awful late and I'm pretty sleepy. I guess you can tell that by the lousy way I'm typing here. I really do have lots of work here to keep me going. This 5th star coming through gave me a lot of headaches, because now I have to make up a whole new roster of who gets five more points, and who that 5 points will give them the 85 required to get them out of the Army. There is so many changes being made all the time that it keeps me going all day and most of the night. I've got a 150 service records that have to be kept straight, and it is up to me to see that these changes get added into their service records and all of that. The biggest headache is making out that payroll. That is done for this month and I wont have to worry about it again till next month so that is a load off my mind. As soon as things get back to normal again this won't be such a bad job. The work is all done sitting down so the only think I am overworking is a bit of brain power. This job gets me out of a lot of inspections and all that useless Army routine stuff that I don't like.

  I hope I get a letter soon telling me that Bert has come home. How about Garth (Boyce)? Did he get down to the place? How about Dean Hildreth, are they going to keep him around for a while or what? **(Bud stated that Dean Hildreth claimed he was shelled shocked and emotionally unstable and was sent back to the States).** The thing that always had me wondering was where in the heck Dean ever got up enough nerve to jump out of a plane... I guess I shouldn't talk that way; maybe someday I'll crack up. I've seen a lot of good guys go nuts when they get up front. Not really nuts. Their nerves just go all shot and they just are not any good. I got kind of nervous one night myself. I was in Belgium at the time. It was one of those cold winter nights around the 25th of January, and I was stringing telephone lines up to the front companies. The Krauts were throwing in those screaming meemies (artillery rounds). They got a screaming sound to them that would make a brass monkey shiver. Anyway one of them came in and sounded like it was going to land right *in my shirt pocket. I hit the ground by a bunch of cows standing along the road. The shell missed my shirt pocket by a ways and when I got up there were 5 dead cows around me. I've had shells land closer than that to me, but none with the sound that those screaming meemies put out* (later that night Bud was in the basement of a building and continued to hear the wounded cows screaming and moaning all night. These sounds gave Bud the shakes and unnerved him to a point he thought he might loose his mind. The Army called this "combat fatigue" Bud just knew he couldn't stop shaking. Bud kept saying to himself I am going to be alright. In the morning he was fine and got back to his duties). *Those Kraut 88's are the next worse sounding thing. Our own 105's are not very pleasant sounding, I'll tell you. The closest I ever came to getting killed was the day after Christmas when the 75th Infantry's artillery threw in some short rounds of those 105's. I was with the Colonel (Battalion Commander Boyle) at the time (see letter from December 29, 1944 for details). I had just gotten a telephone up to him. You should have heard the way he was burning up those wires to get that artillery stopped. After things quieted down he sat there for a half hour and just shook and told me all about the peace time Army. He was quite a guy. One of those big guys that ain't afraid of nothing, but that shell landed practically on us kind of unnerved him a bit. He got all shot up about two weeks later. He wasn't one of those kind of Colonels that sat behind the lines. He was always up there leading the men, but on the 6th of January he walked into an ambush and got all shot up. He got slugs from a burp gun put into him by a Kraut. Nothing could kill that guy though. He is back in the States now and the Battalion hasn't been the same since he left. The guys called him "Wild Bill". He was quite a guy alright.*

The guy that saved the Colonel that night is a guy that I use go to Wilson with (Wilson High School in Long Beach, CA) and he use to go around with Afton. His name is Bob Steel. He got the Krauts that shot up Wild Bill, and then he carried him into the aid station. He got a Sliver Star and was made a second lieutenant soon afterwards. That Bob is a good man too. He has got plenty of nerve.

Well, doggone I really have got to call it a night. It is one-thirty now and I've got to get up at 7:00 am. There is lots I've got to do tomorrow.

Tell everyone hello for me. Thanks again for the swell pictures. If you see Irene again tell her hello for me and that I'll write just as soon as she sends me her new address.

Love
Bud

Letter typed to Mom from Harland "Bud" Curtis

Joigny, France, after the war
June 23, 1945

Dear Mom,

Got a nice long letter from you today, and one from Jill with pictures she has taken out to the house in her bathing suit. Is that the one I gave her? It is kind of late at night as usual. That seems to be the only time I can call my own and get a chance to write to anyone. Well right now I'm sweating out ways and means that I can get out of this darn Army all together. Boy I'm thinking of all kinds of angles, but it is all just chance and nothing definite. You see right now I have 67 points that I can call my own. That purple heart came through and so did a 5th campaign star for Central Europe yesterday (After the war ended the men of the 517th PRCT were instructed to document any wounds they received. Bud had been hit with flying glass from German shells landing close by and was peppered with glass fragments. Other times he suffered cuts from shrapnel, but never applied for the Purple Heart. Now the war was over and he had time to request the medal for wounds he received. He earned 3 Purple Heart medals). *Altogether that gives me 67 points. Now today I turned in for a cluster to my purple heart and the chances are 99 to 100 that it will go through, and if it does, that will give me 72 points. Okay, now what the rest depends upon is how far they will lower the critical score to, and how many more points I can chalk up in the meantime for things that happened before May 12th. Here's the deal on what I might get ten more points on and enough to get me out of the Army if they drop the critical score to 82. When we were in Southern France and along the border line of France and Italy we had some men go on patrols over the border into Italy, which was Northern Italy. That was all that we did, but yet Northern Italy is another campaign, and they are trying to get the star for that which would give me 6 campaign stars and raise me up 5 more points. It is all just chance whether we will get it or not, but I sure hope so.*

Now the other thing is this Combat Infantry Badge that is given to infantry soldiers that have been in combat. There is talk, just talk now or rather rumors that they will give 5 points for that. Well that is what will get me out of the Army, but I can't plan on anything, because nothing is definite. It all depends on whether we get the 6th star for Northern Italy, and whether they will give 5 points for the Combat Infantry Badge. Well if they get these two things through, and this cluster for my purple heart I put in for today comes through that will give me 82 points, and I think there is a good chance that they will lower the score 3 points anyway. The way I feel I would cut somebody's throat to get out of the Army. I can practically count on 5 more points for that purple heart cluster I put in for today, and figure that I've got 72 points sitting right in the palm of my hand even if nothing else goes through at all. Well 72 points ain't nothing to sneeze at. It might even mean that when I get back to the States they might take into consideration those guys that are pretty high up on the points, and not send them out to foreign duty again, but just keep them in the United States. That's just one of those things that I'll have to wait and see what happens, but you never can tell. I have been over seas and have seen plenty of combat, and with 72 points that might mean something when I get back to the States. I'm hoping that it will mean they will at least keep me in the States for good even if I can't get out of the Army (Bud never did get the required points to be sent home. He volunteered to fight the Japanese and was sent back home to California for 30 days furlough. When he arrived home the war with Japan was over. He reported to Fort McArthur for duty. He did not have the required points to be discharged. The Army told him to go home for 45 days and remain in uniform and report back. He did and was discharged on October 18, 1945).

I've got all kinds of different little cuts and that which I got when I was up front, and if I can get away with it I'm going to try and collect on every one of them. I have no pride at all if it will get me out of the Army. I was even going to get a hold of that Lieutenant that put me in for the Bronze Star and ask him to put me in for it again. The last time he put me in for it something went wrong and I never got it. That Lieutenant is in the hospital right now though and there isn't anything that I can do about it. I'm afraid it will be too late when I see him again for him to put me in for it.

I'm going to try and figure out someway to get some more points if I can. Doggone, 13 points standing between me and being a civilian is darn hard to take. Just 13 points and I'd have 85. I've just got to figure out a way to get 10 more points anyway. If that 6th star and this talk of getting points for that Combat Infantry Badge comes through I'll have it made, but there is too much of a chance that something will go wrong there, so if I can possibly do it I'm going to try and get two more oak leaf clusters for that purple heart, but there is a big doubt there too if I can do that. I've already got the purple heart and I put in for a cluster today, so I don't know if I can get these other two or not. I'll try anything though if it will get me out of this Army.

Well tomorrow or rather now for 15 minutes it has been Sunday. I'm getting to try and get to church that the Mormon guys around here have started. I'm going to have to work tomorrow though, but usually around church time they let the guys off to go to church. I've been so darn busy that I haven't had time for anything the past couple of weeks. In just 5 days I'll be taking over this job all to myself, because the corporal is leaving then to go into the band. He is the best piano player and arranger that this outfit has got. He really is good. Well anyway he is leaving and it will be all up to me. If things go right I guess they'll make me a corporal. This shouldn't be too bad of a job after things get back to normal again, but right now I've never had so many headaches in my life. I stepped in here at the busiest time they have ever had and now I'll be taking it all over by myself in just a few days. Well I'll see how it turns out and if I like staying here or not.

I guess I had better go hit the sack. I'm so darn sleepy I hardly know what I'm clicking off here anyway.

Lots of love to everyone. I hope Bert go home.

Bye for now,   Bud

Letter typed to Mom from Harland "Bud" Curtis

Joigny, France, after the war
July 23, 1945

Dear Mom,

I am still busy as ever if not just a little bit more so you will still just have to wait a little bit longer I guess until things quiet down and I can start writing more often. I wrote Jill the other day that in just a couple of days, about the 25th of this month we are pulling out of here and are going to a place I think is called Camp Pittsburg. It is one of those places that you go to, to get ready to make that boat trip again, only this time it will be home. The way things stand now I should be home in August. Better hold up on anything you were going to send me. There might be possibilities that something will come up to keep us here longer, but I don't think so. They are shipping all the guys in this outfit that have 75 points over to the 17th Airborne Division (The 17thAirborne Division was the division the 517th trained with in the United States). *They are going home too, but not until September. Maybe they figure that the point system will be lowered to 75 points, but I don't know right now. Anyway I just missed getting shipped out of here because I've got 72 points. I've got another 5 points coming when I get that Oak Leaf Cluster to my Purple Heart, and I'm just hoping that the orders don't come through on it until after we are well on our way to coming home, or other wise I might get shipped out to another outfit and have to wait longer before I get to come home. For some reason this 13th Airborne Division is getting all priorities on shipping. I guess they figure they'll need them pretty soon over Tokyo way. Well things are kind of uncertain right now, but as the score adds up now it shouldn't be too much longer until I'm walking up the front steps of the house. Maybe even if they lower the points to 75 and I get this cluster I might be coming home to stay. Don't plan on too much though.*

Tell everyone hello for me.

Lots of love   Bud

Letter to Mom from Harland "Bud" Curtis

Camp Pittsburg, Le Harve, France, after the war
July 29, 1945

Dear Mom,

Well here I am in this camp I told you about where they send guys to get ready to come home. The 517th will be on the high seas headed for New York on the 18th of August, but as for me, I'll still be here in France. I'll be shipped out of here in a few days to the 17th Airborne Division. You see mom, I got 77 points and that is too many to stay with this outfit because they are headed for the pacific soon after everyone gets a furlough in the states. Anyway anybody who has 75 or more points goes to the 17th Airborne Division (division patch was the "Eagle's claw") and when they finally get back to the states it will be to stay Eventually, they figure the points will be lowered to 75, but even if they don't I'll still stay in the states for the duration after I get back so that wont be such a bad deal. I sure hate to get this close to that gang plank leading home, but I guess it is all for the best. I don't know when I'll get home now. The Stars and Stripes (the military newspaper) said last week that the 17th Airborne Division was coming home in September but I never believe too much in anything until it happens. Chances are I won't get home until next year.

  The only consolation is that when I do get home it will be to stay. That's better than coming home now and then find myself in the pacific in a couple of months.

  Don't get your hopes up to much for anything but also don't get discouraged either. You see Mom right now I just don't know any real facts about what is going to happen. Some people tell me I'll be home in a couple of months and be a civilian, and other people tell me I'll be sweating out coming home until after December and then when I do get there I might not get out of the Army. It is pretty well settled though that whatever happens I'll never have to go to the pacific, and that sounds good enough for me. When you receive this letter don't write to me until you get a letter telling you what my address will be in the 17th Airborne Division. Even the letters you are writing now might be months getting to me if they chase the 517th around and I ain't there to get them. So just hold up on the mail until I write to you telling you where I'll be. Tell Jill not to write for a while too will you please!!! I hope she hasn't sent those pictures yet.

  They tell me that they censor the mail in the 17th Airborne Division, but for the life of me I cant see any reason for them to censor mail there, and they don't even censor our mail here even, and this outfit is practically ready to sail. I guess it is just whoever is in charge of the 17th has got a lot of screwy ideas. Maybe that is just rumors that mail is censored there, but if it is true I won't be able to tell you things about when I think I will come home.

  Boy, I don't know what to say about Bert going into the Army. I know what I would like to say, but I guess it is too late to do any good now. If he thought the Merchant Marines were giving him a bad deal, just wait until the Army goes to work on him.

  Those things you told me he said about getting the advantage of the G.I. Bill of Rights and going to O.C.S. (officer candidate school) and all that well, all he is doing is just talking himself into thinking he is getting a good deal. What will really happen is it won't be long until he finds himself one of these brand new fresh guys they are talking about that are to relieve the guys with 85 points.

  Maybe you don't realize it but 85 points is a heck of a lot of points if you have to sweat them out for one each month. I was just lucky in getting into 5 different campaigns and getting 15 more points for three minor wounds. Ask the average guy who has only been in the Army for two years how many points has he got, and it won't be very many I'll tell you, unless he was lucky like me. The only thing I hope and pray for is that they won't get him over in the pacific until the fighting is over. I don't ever want him to have to go through the things I have. But even if the war does end he'll no doubt find himself over in Japan as occupation army for a few years. Just wait and mark my words. If he ever gets sent away from home as long as I've been he will never stop hating himself for quitting the Merchant Marines. He has got the best deal in the world if he never got paid a cent, and he doesn't realize it. I wish I could have been there to talk to him. I sure hope he makes OCS alright. It won't be so bad then, but if he doesn't, I can tell you now that $50.00 a month is going to *make him wonder what he was talking about when he said he wasn't getting enough from the Merchant Marines. Well maybe he will do all right. He has lots of qualifications to land him a soft place somewhere, but the Army has a strange way of not recognizing a good man, and he might get tossed around until he lands up in the plain dog life infantry. Whatever happens keep him away from the infantry. It is the infantry that gets killed. If he ever gets in the infantry I won't be able to sleep nights. I know what kind of a life that will be ahead of him. Hey, I don't want to get you all worried now. He will do okay, but just keep him away from the infantry, and you will know for sure that he'll be coming home. I'm afraid that it will be a long long time before he will be putting on civilian clothes again though no matter what part of the Army he is in. Let me know how things turn out. There is a lot of advice I could give him but that wouldn't any good. He'll just have to learn the Army ways the hard way. The one and only thing I can say that will do him any good if he listens to me is "just keep away from the infantry." No mater what other part of the Army he gets into he can be sure of coming back home again in one piece. There isn't anyway that can keep him from not staying in the Army for a long time now, but there is lots of things a guy can do to make life in the Army better for himself if he does the right thing at the right time. If you really want to know what I think about Bert going in the Army I'll tell you. As much as I hate this Army and want to get out, and now I am just about out I would gladly if possible take Bert's place now and go through it all over again just to keep him away from the hell I have already been through, and I would go through it all again to keep Bert home if I could. That is what I think about him going into the Army. He doesn't realize what he is getting into, but I do, and I would much rather that I could take his place and do it again just to keep him from having to do it once. I'll be home before he ever leaves the states. I'll get to see him. Enough for now* (Bert never had to go into the Army, but remained in the Merchant Marines. He traveled to Japan taking needed supplies after the war ended. He was an Electrical Officer and in charge of the entire ship's electrical systems. As an officer he had special privileges, and one was he was allowed to take his Harley Davidson motorcycle on board. When he arrived in Japan he rode his motorcycle where ever he wanted to. Before leaving Japan, he sold the motorcycle for twice it's worth to a Japanese person. Bert returned to Long Beach, California, in 1946. He resigned from the Merchant Marines in 1947, and began a career as an electrical contractor).

Lots of love

Bud

In Bud's letters he was incorrect about the point system. He actually had 82 points, but was not eligible for transfer to the 17th Airborne Division. Soldiers had to have at least 85 points to be eligible to be reassigned to the 17th Airborne Division to be shipped home to the United States for discharge. The 17th Airborne Division served as a transfer unit for men eligible for discharge. However the war with Japan was still going on and Bud had volunteered for combat duty to fight in the Pacific. Bud and the others were sure the 517th would go fight in the pacific, but now that was changed. The 13th Airborne Division was ordered to go. Webster (2002) reported, "The 13th Airborne Division was starting to leave Europe for the invasion of Japan by way of the United States." In what seemed like seconds, Bud was reassigned from the 517th PRCT to the 13th Airborne Division (unit patch was the "Unicorn"). He was now out of his beloved outfit and with men who had never seen combat. The 13th Airborne Division arrived in France a few weeks before the war ended in Europe, and never saw combat. To the Army it made sense to assign them to go fight in the pacific. Bud also learned he could be assigned with the 82nd Airborne Division that was given Occupational duty in Berlin. He wanted nothing to do with occupation duty and was glad he had volunteered to go fight in the Pacific. He was promised 30 days furlough upon returning to the United States from Europe in route to fight in the pacific. It looked like he was on his way to fight in another war.

Bud left France around August 12, 1945, and sailed on the Oneida-Victory from Le Harve, France. He was out to sea only 3 days when they were notified that the war in Japan had ended on August 16, 1945. Rumors began on the ship that it would turn around, and head back to France so that soldiers in Europe with 85 points could be sent home on the ship. This would mean that Bud would have to stay in France until he could attain the 85 points.

This rumor lasted about a half a day when they were told the ship would not turn around, and within a week they arrived in New York Harbor. Bud took many picture of his arrival. They came down the Hudson River, past the Statue of Liberty, with fireboats spraying and shooting water into the air. Ship horns were blasting a welcome to these returning troops. Along the shoreline were banners saying, "Welcome home, job well done." When Bud got off of the ship he was taken to an Army camp in the New York City area. He stayed there for 3 days being processed, and waited for train transportation to Fort MacArthur, California. Bud spent 5 long days on the train returning to his hometown. When he arrived at Fort MacArthur he was taken to the out-processing center and his discharge papers were filled out. When he arrived home there were very few soldiers roaming the streets. He perhaps was the only Army Paratrooper back in Long Beach, California at that time. What stories he would have to tell. His nightmare of combat was over.

Howard Hensleigh, Second Battalion S-2 and an expert on the 517th PRCT reported on March 5, 2009, the leadership of the Combat Team from start to finish. He said,

> "The 517th Parachute Infantry Regiment was part of the 17th Division until shortly after Tennessee maneuvers, when we became a regimental combat team by adding the 460th Parachute Field Artillery Battalion and the 596th Parachute Engineer Company. We did all our fighting in Italy, France Belgium and Germany as a combat team. In the Spring of 1945 the combat team was broken up and the 517th became a parachute infantry regiment of the 13th Airborne Division, with the 460th and the 596th being absorbed into the 13th's artillery and engineering units. We were on our way through the States for the invasion of Japan when the war ended as we crossed the Atlantic.
>
> Our original regimental commander was Louis Walsh, a 32 year old WestPoint graduate who was promoted to colonel soon after taking command, but replaced by Lt Col. Rupert Graves, who graduated from WestPoint about 10 years before Walsh did, and also was promoted to colonel in that job. The regimental executive officer was Lt. Col Walton, who broke a leg on jump into Southern France.
>
> The three infantry battalions were commanded by: 1st Battalion, Bill Boyle; 2nd Battalion. Dick Seitz and 3rd Battalion. Melvin Zais, all were in their mid 20s. When Walton broke his leg, Zais became the combat team's executive officer and Forrest Paxton, who had been regimental S-3, commanded the 3rd Battalion, until he left on points in the Spring of 1945. Tom Cross, who had been executive officer of the 2nd Battalion, commanded the 3rd Battalion. until the 517th was disbanded in 1946. Lt. Col. Cato commanded the 460th and Bob Dalrymple commanded the 596th.
>
> Many of the officers and enlisted men continued in the Army after the war with illustrious careers. Dick Seitz and Mel Zais ended up respectively as three and four star generals. Many who entered civilian life kept one boot in the Army as reservists and National Guardsmen."

> 13 June 1945
> Date
>
> I certify that I have read and understand Redeployment and Readjustment Memorandum Number 7, Headquarters, XVIII Corps, dated 28 May 1945 and I hereby volunteer for assignment to a Category II unit. I fully understand that a Category II unit is one which is scheduled for redeployment to an active theater of war. I further understand that this decision is irrevocable.
>
> Signature *Harland L. Curtis*
> Pfc. 39292962

The form Bud signed to volunteer to go and fight in the Pacific Theater, June 13, 1945

Chapter 15
Home and Readjustment to Civilian Life

When Bud arrived back in California he found the hits of 1945 were: On the Atchison Topeka and the Santa Fe, by Johnny Mercer. My Dreams Are Getting Better All the Time, and Sentimental Journey, by Les Brown, It's Been a Long Long Time, by Harry James. If I Loved You, and Till the End of Time, by Perry Como, Gotta Be This or That, by Benny Goodman. Don't Fence Me In, by Bing Crosby

Bud found out right away that Fort MacArthur was not equipped for the many returning soldiers to keep all of them busy with military duties. Bud needed 85 points to be discharged. He had 82 points. The Army had nothing for Bud to do so they told him to go home for the next 45 days. That would be enough time to accumulate 3 more points for the needed 85 points for discharge. During the next 45 days Bud didn't let any grass grow underneath his feet. He immediately enrolled at Long Beach City College, and started his education. He was required to wear his uniform during this time because he was still on active duty. Many people were impressed with all of his ribbons, decorations, and shining paratroop boots. He was one of the first soldiers to arrive home in Long Beach after the war ended. All the girls wanted to know about the ribbons, decorations on his uniform, and what he had done in the war. He had a lot to tell!

When the 45 days ended Bud reported back to Fort MacArthur, and on October 18, 1945, Bud received his Honorable Discharge from the United States Army. He finally got the 85 points needed for discharge. Prior to leaving Fort MacArthur, the Army tried to talk Bud into staying in the Army Reserves. He was told of the extra money he would earn, promotions, and a chance for retirement from the military. All of this fell on deaf ears. Bud had served his country well, but he had all of the Army one person could take. Bud respectfully declined to stay in the reserves. This proved very beneficial for Bud when in 1951, the Korean War broke out, and every reserve unit was called to active duty to fight in Korea.

Now it was time for the rest of his life to begin. He had survived the war! He could have been in any branch of the service. In fact he was in the Navy for about five minutes, but decided he had to be with the best. Paratroopers, he knew they were elite and would see the toughest fighting of the war in Europe. He had survived combat in Italy, parachuting behind German lines in Southern France, the Battle of the Bulge in Belgium, and the final march into Germany. Twenty Million men died in World War II, but Bud survived. He was one of the lucky ones. Now he had his whole life in front of him, but how would he adjust to civilian life? For the past two years officers, sergeants and corporals gave him orders. His life was controlled and organized. He knew from his paratrooper training he would be successful, and have the confidence to succeed. However, he needed time to decompress from his experiences. He lived at home with his parents, worked with his brother and father learning the trade of an electrician. The stress of war took its toll on Bud, and his relationship with his boyhood sweetheart Jill. Bud was just a young man, and perhaps still just a boy in 1944 when he left for combat duty. The combat and horrors he saw changed him forever, and he was no long a boy but a man; a combat hardened man who had witnessed first hand the most horrific combat of the European war. If not for Jill, Bud may not have survived the war. She gave him purpose to live. Little did Jill know she more than likely saved Bud's life. Her willingness to be his girlfriend, and write letters while he served in the Army gave him hope to continue, and a reason to want to make it home. However, he did not come home the same kid that Jill knew. He was different. Soon after he came home their relationship ended. The reasons are not important, but it is sufficient to say that both had out grown each other and their lives would take different directions forever.

Bud continued at Long Beach Community College taking stenograph classes. Bud's limited training as an office clerk in the Army spurred him on to want a career in business management. Stenograph training seemed to be a natural way to begin. Bud learned short hand and how to operate a stenograph machine. He was going into court reporting.

Bud did everything to keep himself busy. Working with his brother in the daytime as an electrician, and going to school at night. He even took flight training at the Long Beach Airport mostly on the weekends learning how to fly a plane. He still craved adventure. He had the G.I. Bill for education and he was using it. Keeping himself busy still did not keep him from remembering the horrors of war. He found himself waking in the middle of the night in a cold sweat thinking he was back in battle. Some nights it was too much to bear, but he dealt with it. One night at the dinner table with his parents, a loud bang rang out!

Before anyone could think about what happened Bud was under the table taking cover. His mother said he was never the same person after the war. The war had changed him forever. He would suffer posttraumatic stress for the rest of his life.

Bud's life seemed to be going nowhere. He was discharged in 1945 and now it was 1947. One night he went downtown on the oceanfront near the "Pike" (The Pike was an amusement park where Bud went as a kid). With some friends they went to many of the nightclubs that lined Ocean Boulevard. Just off of Ocean Boulevard and down by the Pike was one nightclub that Bud went into that would change his life forever. As he came through the door he looked up at the stage and saw the most beautiful girl he had ever seen. She was the featured singer and musician in the band that was entertaining there. He was smitten. Being a fearless paratrooper he immediately went up to her and introduced himself. Bud wanted a date right away, but she said no. That didn't stop him and he kept coming back to see her. She finally gave in and agreed to date him. Bud began dating Lois Stenquist and their courtship lasted a few months and on June 13, 1947, Bud and Lois married. Bud's life finally had direction, but the newlyweds were broke. They would have to move into Bud's parent's house. Bud's brother, his wife and son lived there also. It was not easy for any of them to be under one roof. In fact parents Bert and Luciel moved out into an apartment so the newlywed couples could have more space. Bud was out of work and needed a job badly. His brother Bert had started his own electrical contractor business and offered Bud a job as an apprentice electrician. Bud quickly accepted the offer and thought now he would be like his father and brother, an electrician. One day at work Bud was on a ladder when he slipped off the rung ripping his arm open. He was rushed to the hospital and required many stitches to sew up the wound. He knew he couldn't make it as an electrician. He had to find another type of work.

Bud soon found work with Lincoln – Mercury of the Ford Motor Company. He became a district service engineer and resolved technical and mechanical problems for customers who were unhappy with their servicing dealerships. Bud was very successful in this job, and was moving up the corporate ladder, but not fast enough. In 1958, Bud accepted a position with "Renault" as the West Coast Operations Manager. He now was responsible for supervising the arrival of every new Renault coming from France into the western United States.

Lois Stenquist and Bud Curtis on their wedding day. Lois standing in front of Bud's parents home after their wedding.

With these experiences he became a National Import Operation Manager for Mercedes Benz, and in later years other major automobile companies. Bud would remain in the car industry for the rest of his working life. His children grew up respecting the military, and knowing what it meant to be an American. They in turn taught their children what their grandfather did in World War II. Their children are now teaching the great-grandchildren of their great-grandfather and the 517th PRCT. After 60 and half years of marriage, Bud and Lois reflected upon what they accomplished during their lives together. They are the very proud parents of three sons who gave them seven grandchildren: four grandsons, and three granddaughters. Then twelve great-grandchildren: seven boys and five girls. The legacy of Army paratroopers will never be forgotten in the Curtis families because Bud wrote these wonderful letters. We will always remember his war experiences and the men of the 517th Parachute Regimental Combat Team.

The only picture of Bud with a mustache in his Class A Uniform after the War in 1945. He is wearing his Jump Wings, Good Conduct, and European Theather ribbons. Below his ribbons is his Combat Infantry Badge

EPILOGUE

Honored in France
60 Years Later

    It was now noon Paris time, June 4, 2004, when the Air France stewardess announced for the passengers to return their seats to the upright and locked position. Bud's mind quickly came back to the present day. He was flying to Paris to be honored for his military service by the French Government. Where had all of the years gone? It had all happen way too fast. As Bud looked out the airplane window he couldn't believe what he was seeing. The French government had rolled out the "red carpet" for a hero's welcome to the returning American soldiers. With him to witness this honor that France was about to bestow were two of his sons. After the many speeches and praising the American war veterans at the airport they were all taken to various plush hotels in the Paris area. Bud and his sons rested there for the next day's activities. At 10:00 A.M. Bud and the other 99 World War II veterans were taken to the Hotel Des Invaldes (not a hotel, but the nation's memorial for war veterans and where Napoleon is buried). There a French military ceremony was conducted and each veteran was presented with the French Medal of Honor, the "Legion of Honor."

How did this great honor come about? Why was Bud so honored to be one of only 100 World War II veterans to be flown to France? This is the story of his trip to France.

It all started when Bud came to visit his oldest son in Salt Lake City in April 2004. This visit was very important to both Bud and his son because just six months earlier Bud and two of his sons journeyed to Toccoa, Georgia, and Fort Benning where Bud told them all about his paratrooper training. Now many more questions were surfacing, and L. Vaughn had his letters. They spent many hours discussing their trip to the Toccoa reunion in September of 2003, and his World War II experiences. Finally after all of these years he was able to discuss some of the things that had happened to him while he served in the 517th PRCT. During the many conversations Bud began discussing the awards his unit (517th Parachute Regimental Combat Team PRCT) received during the war. The French government had awarded the First Battalion of the 517th PRCT the French Croix de Guerre for fighting the Germans in France. Bud discussed the honor the French had given the First Battalion in his letter to his father, dated June 22, 1945. He told him, "The French presented us with some kind of decoration. I forget just what it was, but it entitles us to wear a braid of some color that goes around the left shoulder."

Being unsure if this distinction entitled the men of the first battalion to wear the French Croix de Guerre medal, I decided to do some research. On April 8, 2004, I told my Dad I was going to find out if he was authorized the medal, and how I could get a French Croix de Guerre medal for my Dad. I first telephone the French Embassy in New York City. They didn't know anything about it, and told me to call the French Consulate in New York. They also did not know about the medal, and suggested I telephone the French Embassy in Washington, D.C. I was beginning to think I might not find the answer to this question until I spoke with a wonderful Frenchman, by the name of Thierry Fiorini. He was very helpful, and seemed to know all about the Croix de Guerre Medal. As we became better acquainted I found out he was a Sergeant Major in the French Army, and I believe was the senior enlisted man at the Embassy. The Sergeant Major told me that individual soldiers of the 517th PRCT were not eligible for the medal since it was given as a unit award, and they were only authorized to wear the cord around the left shoulder. This made sense to me, and I started to thank him for his time when he said however, the French government is awarding 100 Legion of Honor Medals to World War II soldiers who fought in France.

I asked him how my Dad would be eligible to receive this award. He informed me that I had to contact a U.S. Veteran's Administration (VA) representative in Washington, D.C. I asked for his name and phone number, which he gave me. I telephoned Mr. Robert Elliott, VA representative, and asked him about my Dad receiving the medal (I thought it would be sent to him in the mail).

Mr Elliott told me the 100 candidates to receive the Legion of Honor were already selected. He told me he was in the final stages of submitting their names to the French Government. He said my Dad could be put on a standby list in case one of the recipients could not go. He informed me that the oldest man going was 94 and the youngest was 78. He told me that the award would be presented to 99 of 100 World War II recipients in Paris France on June 5th, 2004, at the Hotel des Invalides, and one man would receive the medal at a special ceremony at Arromanches les Bains (Sword Beach) on June 6, 2004, by French President Chirac (I later found out the man to receive the medal was from San

Diego, served as an adjunct professor at San Diego State University. This man landed on the beaches in the very early hours of June 6th, and went to the homes French sympathizers that support the German Army. There he held a gun to their heads and made them broadcast false messages to the Germans to confuse them during the D-Day invasion, June 6, 1944).

In our telephone conversation Mr. Elliott explained that each recipient had to be in good shape to make the trip and could not be in a wheelchair. He said these recipients would be required to do a great deal of walking. Mr. Elliott, then gave me instructions as to what he wanted for documentation, and instructed me to mail the application to him at his home address as soon as possible. I quickly prepared a packet for my Dad, sending this VA representative a copy of Dad's discharge papers, a chronological outline of the 517th Parachute Regimental Combat Team (PRCT) war record, especially their service in France. I provided documentation to Mr. Elliott that the 517th was the only unit in World War II history to spend 94 consecutive days on the front lines without relief. These days started when the unit jumped into southern France on August 15, 1944 and continued for 94 continuous days. I also sent two pictures of my Dad in full combat gear that has appear in this book and many other books (see page 162). I knew time was short, and even if Dad was selected as one of the recipients he may not be able to obtain his passports in time. I quickly mailed off the packet express mail and anxiously awaited his reply.

Two days later on Monday, April 12, 2004, much to my surprise Mr. Elliott telephoned informing me he had received the application packet, and was very impressed with what I had mailed him. He then told me one of the original recipients could not make the trip to France due to health reasons. His next statement shocked me, and I could not believe what I was hearing. He said, "Bud Curtis was now moved up from "standby" to being one of the 100 men to be presented the Legion of Honor Medal in Paris, France, on June 5th 2004." He also told me my father could have one escort with him all expenses paid. I thanked Mr. Elliott for his kindness and for selecting my father. I quickly called Dad and told him to get his bags packed, and to get his passport renewed. Most men were taking their wives with them, but Mom was unable to make such a long trip due to health reasons. I became the second luckiest guy in the world as Dad asked me to go with him as his escort.

I telephoned Mr. Elliott and told him Dad had asked me to escort him to France. Mr. Elliott told me other members of the family could come to France to see their Veteran received the Legion of Honor Medal, but would have to pay their own expenses. Mr. Elliott also said that the 100 veterans with their escorts, and family members would be taken to the Normandy beaches on June 6, 2004. There they would sit in front of Presidents George W. Bush and Jacques Chirac, and listen to their addresses about the invasion of France, "D-Day, June 6, 1944.

I still had a hard time believing we were really going to France until I saw this article on line in the Baltimore Sun on April 21, 2004:

300 D-Day Vets Invited To Paris For Anniversary

PARIS - The French government announced plans yesterday to welcome 300 veterans of the 1944 D-Day landings - chiefly from Britain, Canada and the United States - with free accommodation in top Paris hotels.

Ceremonies will be conducted in early June to mark the 60th anniversary of the Allied landings on the beaches of Normandy in northwest France. The invasion began June 6, 1944, and was a major turning point in World War II to end Nazi rule.

Former servicemen are expected from 13 countries and will be decorated with the Legion of Honor at a ceremony at the Invalides on June 5, 2004.

On June 2, 2004, I left Salt Lake City, Utah, and flew to Long Beach, California. There I met Dad, and we flew to the Washington, D.C. Dulles Airport. Mr. Elliott had made arrangements for all veterans to stay at the Fairfield Inn, Chantilly, and at about 7:30 P.M. we checked in and received our security badges from Robert Elliott. At about 9:00 P.M. Dad and I drove the 25 miles to downtown Washington, D.C. to see the new World War II memorial that was just dedicated a few days earlier on Monday, May 31, 2004, by President Bush.

The next day June, 3, 2004, we were shuttled to the Airport Marriott Hotel where we sat in a banquet room until about 6:00 P.M. One of the veterans came up to us and saw my 517th PRCT shirt and said, "I was in the 517th." Dad immediately recognized this man to be Walter Goforth. Walter and Bud had served together in the Communications Section of Regimental Headquarters while at Camp Mackall. Dad and Walter visited for hours talking about their experiences in the 517th. Bud found out that after the war Walter went to Medical school and became a doctor. What a great accomplishment.

Bud and Dr. Walter Goforth at the Dulles Marriott Hotel, June 3, 2004.
Bud Curtis revisits the World War II Memorial on June 30, 2007.

At the Marriott Hotel the veterans mingled and exchanged stories of their war years. They ate plenty of snacks, and at about 6:00 P.M. all of us were taken to the French Embassy in Washington, D.C. where many speeches were made by French dignitaries. The dignitaries especially the French Ambassador spoke to the veterans, and told then how appreciative the French people were of what the World War II veterans did for their country. They said, "France would never forget." After these speeches they served drinks and French pastries. At about 8:00 P.M. they took us by bus back to the Dulles airport and there we flew on a special charter Air France flight to Paris.

French Embassy in Washington, D. C. Bud being interviewed by a French reporter, June 3, 2004

Bud on the Air France Jet traveling to Paris, June 3, 2004

The flight to France left Washington, D.C. at 10:00 P.M. and flew for about 7 ½ hours to Paris. We arrived in Paris at noontime on June 4, 2004. When Bud looked out the aircraft window he was totally surprised when he saw a red carpet with a military band playing and reporters waiting to take pictures. It was a great welcome. We were not expecting this great treatment. We were escorted into a building where French dignitaries gave speeches along with the American Ambassador to France giving a speech. Once again the French said, "We will never forget." I was beginning to see the sincerity of these people.

A "Red Carpet" Welcome at the Paris Airport, June 4, 2004 Next we were taken to the 4 Seasons, George V Hotel, (a five star hotel). It was very nice and elegant. There we met Tim, Bud's second son who had flown to France earlier that day. There was a reception where Bud spoke with the former Prime Minister of Canada.

Bud Curtis discussing the issues of the day with the former Prime Minister of Canada June 4, 2004

Bud outside the Hotel des Invalides just prior to receiving his medal, June 5, 2004

Mr. Michel Barnier, Minister of Foreign Affairs pinning on Dad's Legion of Honor June 5, 2004, at about 11:00 A.M.

After the medal was presented we were taken to a reception hosted by the French American Association where a luncheon was served. I asked Tim to eat lunch with Dad and be part of these festivities.

The next morning, June 6th we had to be on buses at 4:30 A.M. Bud put on his suite coat that had his newly presented Legion of Honor and his other U.S. medals. The buses transported us to Gare de-Saint-Lazare which is the train station. Security was tight with police officers everywhere. We went through metal detectors, and the veterans were escorted to first class accommodations on the train.

We arrived in Caen France at about 8:00 am and were taken by buses to Collierville, France, which is where the American cemetery is located on Omaha Beaches. We had police officers on motorcycles with sirens blaring when necessary escorting us through the traffic. Other vehicles were required to stop while we were allowed to pass. It really was VIP treatment. We arrive at the American Cemetery, and there we had to walk the long path, maybe a ¼ mile to the reviewing stand. We saw Omaha beach where the Americans had to come ashore on that fateful day, June 6, 1944. The Germans had a commanding view and as history has shown was a very deadly place for Americans.

Bud and I were escorted to a seating area just in front of where the presidents were to speak. We were on the 5th row in front of the Presidents. We saw General James Jones, USMC, Supreme Allied Commander, General Edward Meyer, Chairman of the Joint Chiefs of Staff, Secretary of State, Colin Powell, Mrs. Bush, and Condoleezza Rice (the current Secretary of State). The weather was warm and the sun was out which was unusual for the Normandy area. We saw President Bush's helicopter fly in and land behind some trees. The ceremony started with Presidents Chirac and Bush laying a wreath at the war memorial monument. It was so far away we could not see what they were doing. They walked down a long red carpet to the reviewing stand. When they arrived, the French National anthem was played, then the American national anthem. When this was played the World War II veterans began singing. President Bush didn't really know what to do. He started shifting his eyes back and forth, and then began to mouth the words without really singing. After the anthems President Chirac first spoke. We had headsets to listen to President Chirac's speech in English since he spoke in French. Next President Bush spoke.

Presidents Chirac and Bush giving their speeches at the American Cemetery at Omaha Beach

Once the ceremony at Omaha Beach was over we got to walk around the cemetery for a while and look at the beach. Then we boarded our buses and were whisked away to our next stop. Arromanches les Bains. This is where Sword Beach was when the British landed on D-Day. We were given a lunch in a nice carrying case, and then escorted to bleachers to sit for over two hours listening to the many speeches given by the heads of state.

Omaha Beach

Left photo of Queen Elizabeth in blue dress and hat at ceremonies at Arromanches les Bains, Sword Beach, June 6, 2004. President and Mrs. Bush arriving in photo at right

      Once the ceremonies were over, about 4:00 P.M., at Arromaches les Bains, we waited for an hour before boarding our buses. The buses finally came and we were then returned to the train station in Caen. We boarded the train for a two hour ride back to Paris; then a bus ride to the hotel where we arrived at 1:00 A.M. Monday morning, June 7, 2004. It was a very long day and all three of us were BEAT! (extremely tired)

Bud at the park near the Eiffel Tower on June 7, 2004

On Monday, June 7, 2004, we toured Paris. Then on June 8, 2004, we left Paris for Nice France to visit Bud's World War II combat sites. We went by "Bullet" train leaving from Garde de Lyon, which travels south. The trip took five and ½ hours to Nice. The countryside reminded me of parts of Utah and California. It was beautiful. We rented a mini van, and the next day, Wednesday, June 9, 2004, we began our adventure to find Dad's old fighting grounds from World War II. We began driving up the narrow mountain roads, and we came to a road sign that said Sospel. We knew we were in the right area. This is where Bud had fought 60 years earlier. As we arrived in Sospel, we found a beautiful little small French village and there Bud told us his stories of that area.

Dad at the crossroads of Sospel and Sospel.
We were not sure which way to go, but on June 9, 2004, we
took the road to the left and got to Sospel.

Sospel, France, June 9, 2004

We left Sospel after lunch and drove to Col des Braus. This is where Dad lived in a mountain tunnel for six weeks. Tim and I were very excited to see this place. There were many letters Bud wrote to his mother about living in this tunnel. When we got to Col des Braus we expected to see many shops like in Sospel. There was nothing except one house and a small restaurant. Bud did not recognize the area. We drove down the hill and up another before we realized we had gone too far. We turned around and drove back. From this view Bud recognized the area immediately and we found the tunnel. We drove the rental car right to the tunnel. Dad went up and touched the tunnel door and tried to open it, but it was welded shut. He told us stories about the area and the shelling that went on between the Americans and Germans. He told us about the communication section's cat, Adolph, who whiskers were long, and reminded them of the black mustache of Hitler.

Adolph was a valuable cat because he could hear incoming shells before the troopers could. When Adolph ran, you got in the tunnel fast because German shells were coming in.

Tim and I began looking around on the ground. We found lots of shrapnel from German artillery rounds, sixty years later. Across the street was a partial building that we discovered was the one our Dad stayed in his first night in Col des Braus. He told us there was continuous shelling coming in from the Germans. In Bud's letters home to his mother in 1944 he talked about being under a bed in a house in Col des Braus while German shells were landing just outside. This partial building was that house Bud was in.

Pictures near the Tunnel, and Adolph the cat taken in 1944

Home in Col des Braus in which Bud stayed his first night while Germans shelled the area. The picture on the right is how the building looked on June 9, 2004

Bud at the Tunnel, 60 years later in Col des Braus, June 9, 2004

Phillip, the proprietor, and Bud at the restaurant in Col des Braus, and outside the restaurant

Bud in Col de Braus with a view of the only existing house, restaurant, June 9, 2004

Next we met with Phillip the proprietor of the only restaurant in Col des Braus. When he found out Bud was from the 517th PRCT he became very exited and welcomed us. He had plaques on the wall dedicated to the 517th PRCT. We reluctantly left Col de Braus and proceeded to Peira-Cava, a town high up in the French mountains.

Standing by the road signs to places of interest, and Bud in Peira-Cava

Then the long drive from Peira-Cava back to Monaco. The next day June 10, 2004, came early and we drove the hour and half to Le Muy. When we arrived we saw a sign directing us to a museum. There much to our surprise the museum was dedicated to the 517th and the other units that made the jump into France on August 15, 1944. While touring the museum we met a couple from England, Mr. and Mrs. Folkard. Mrs. Folkard's father was a British paratrooper who made the jump with the 517th. She explained they were very close friends with the curator, Michel Soldi. Mr. and Mrs. Folkard then took us to a memorial site where it honored the paratroopers who landed that morning. Our plans included visiting the National Cemetery in Draguignan. The Fokard's told us they would set up an appointment to meet with Michel Soldi upon our return from the cemetery. When we returned to the museum Michel was there a news photographer and took pictures of Bud for their newspaper.

Musee de Liberation in Le Muy, France, June 10, 2004, Bud with sons Lory and Tim

Bud at the American Military Cemetery in Draginan, France.

Michel Soldi, and Eric Curators of Musee de Liberation, Le Muy, and Bud with his sons

The general area where Bud landed in his combat parachute
jump into southern France, August 15, 1944

  As we left the monument in Le Muy, and quickly drove to the American Cemetery in Draguignan we saw the graves of American soldiers who died during the invasion of southern France. We then drove back for a 2:30 P.M. appointment with Michel, Eric, and the news photographer. Our time by then was far spent, and we had to get back to Nice to catch the plane to Paris. Michel took us to the spot where he knew the first Battalion of the 517th landed. Michel bid us farewell, and we drove back to Nice, turned in the rental car, checked in with Easy Jet Airlines, and flew to Paris.

Easy Jet flight from Nice to Paris France, June 10, 2004

We arrived in Paris around 7:00 P.M. and took a taxi to our hotel. The next day June 11, 2004, we flew from Paris to Los Angeles. It was 10 hours, 43 minutes and 27 seconds. A very long flight.

The ceremonies and the trip to southern France were now over. The memories of being with our father to revisit the battlegrounds of Le Muy, Col de Braus, Sospel, and Para Cava would never be forgotten. Bud had fought for the freedom of the world. He took his two sons with him to France. There he told the stories of his war experiences. His sons marveled at the sacrifice he, and other soldiers had to make to keep this world free. Now sixty years later the French expressed their gratitude for him doing his duty as an American soldier. Somehow it all seemed worth it to Bud. Although he and others never wanted to go to war, they did. They performed their duty without complaint never asking for any special treatment or reward. They did what they had to do. Their country called, and they responded. The men who volunteered to be the 517th Parachute Regimental Combat Team (Battling Buzzards) during World War II put their country above their own lives.

When the war ended and the accomplishments of the 517th were compiled, it was determined they fought in five campaigns on battlefields from Italy to France, Belgium and finally Germany. The 517th fought ninety four continuous days on the front lines in France without relief, more than any other unit in World War II. No other unit can claim this distinction. They suffered 1,576 casualties and had 247 men killed in action. The soldiers of the 517th Parachute Regimental Combat Team earned one Medal of Honor, 6 Distinguished Service Crosses, 131 Silver Stars, 631 Bronze Stars for Valor, 1,576 Purple Hearts. 5 Legion of Merits, and 17 French Croix de Guerres (and yes two Legion of Honor Medals, Bud and Walter, June 5, 2005). The 517th PRCT was one of the U.S. Army's first elite airborne combat units. They endured some of the heaviest fighting in the European campaigns, and produced an astonishing eight generals, with four troopers starting as privates.

This was truly the "Greatest Generation," the best America could offer. When the war was over these men resumed a normal life never asking for any special treatment. Now America, France and the rest of the world were finally giving the World War II veterans the recognition they so deserved. For sixty long years Bud never would talk about his war experiences. Now his stories are recorded for his posterity, and the world. After eighty years he could finally start to talk about his war experiences. Each of us must never forget what World War II veterans gave to this country and the world. We must never let such a world calamity happen again. Bud's and other veterans' services must never be forgotten. The world has a responsibility to remember what these men did to keep the world free.

*Ambassade de France
aux États-Unis*

L'Ambassadeur

Washington, May 06, 2004

Dear Mr. Curtis,

It is my great pleasure and a great honor to inform you that, by decision of the President of the French Republic, you have been chosen to be named Knight of the Legion of Honor, French most prestigious award.

This prestigious French distinction is conferred on you by the French Government in recognition for your participation in the liberation of France during the Second World War.

You are part of one hundred American veterans who will be decorated in France on the occasion of the 60th anniversary of the Normandy landing. On this occasion, you will be the guests of the French people and will be invited to attend the celebrations of that historic event.

A special flight will leave from Washington DC on June 3rd in the evening and we very much hope you will be able to participate. Just before, at 4 pm the 3rd of June, it is my pleasure to cordially invite you and your guest to attend a reception at the French Embassy, 4101 Reservoir Road, Washington DC that will be held in your honor, to pay tribute to your outstanding actions.

You will find here under, one annex informing you of the first practical details of this trip to France. To allow my services to organize as well as possible your stay in France, I would be very grateful if you agree to send back the attached questionnaire. My services will be very happy to answer any of your further questions.

Again, it is a great pleasure for me to convey to you my most sincere and warmest congratulations.

With best regards,

Sincerely,

Jean-David Levitte

Letter received by Bud Curtis, May 6, 2004, from the French Ambassador to the United States inviting him to France to become a Knight of the Legion of Honor, France's most prestigious award.

APPENDIX A

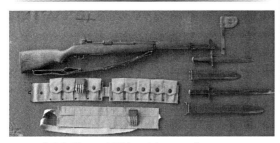

M-1 Garand Rife with muzzle cover, standard M-1, 10 inch bayonet & scabbard, with earlier M-1905, 16 inch bayonet & scabbard discontinued as war proceeded. M1936 cartridge belt with Bandolier with 30.06 ammunition and clip. Equipment courtesy of American Patrol Company Collections, owner Raymond Meldrum.

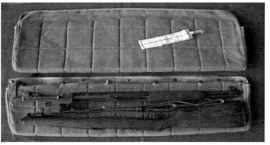

M-1 Rifle disassembled on top of Griswold Bag carried by paratroopers during a jump. Photo of Bud on Book cover shows him standing with Griswold Bag. Bud qualified Expert with the M-1. Rife and equipment courtesy of American Patrol Company Collections, owner Raymond Meldrum.

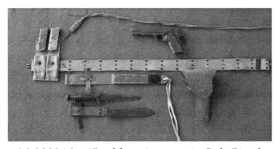

M-1911A1,, 45 caliber Automatic Colt Pistol (ACP) with two magazines, pouch, lanyard, M1936 Pistol belt, M-3 fighting knife and leather and fiberglass scabbards. Equipment courtesy of American Patrol Company Collections, owner Raymond Meldrum.

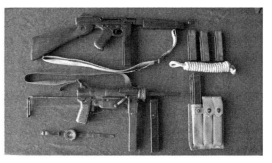

M-1 Thompson Sub Machine Gun (SMG) with magazines (featured in the movie "Saints and Soldiers" with paratrooper rope. M3A1 Sub Machine Gun commonly called the "Grease Gun" with magazines and paratrooper wrist compass. Weapons and equipment courtesy of Alan Firestone.

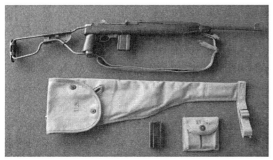

M-1A1 Folding Stock Carbine with case, magazine, and pouch. Bud qualified Sharpshooter with this weapon and jumped into combat with it on August 15, 1944. Rife and equipment courtesy of American Patrol Company Collections, owner Raymond Meldrum.

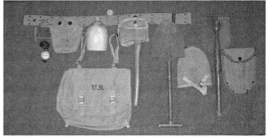

M1936 Pistol belt, lensatic compass with pouch, canteen with cover, M1936 mussett bag, hatchet with carry case, old M-1910 entrenching tool with case commonly called the T-handle shovel. M1943 entrenching tool with case. Equipment courtesy of American Patrol Company Collections, owner Raymond Meldrum.

M1919A4 Browning .30 Caliber Air Cooled Machine Gun. Bud Curtis qualified Expert with this weapon. Weapon courtesy of American Patrol Company Collections, owner Raymond Meldrum.

M1918 Browning Automatic Rifle (B.A.R.) used by soldiers and paratroopers during the War. Weapon courtesy of American Patrol Company Collections, owner Raymond Meldrum.

M2 Paratrooper Helmet in foreground with solid wired rounded chin strap connector and specialized airborne suspenders. Typical ground infantry helmet in background with swivel chin strap connector. Helmets courtesy of American Patrol Company Collections, owner Raymond Meldrum.

REFERENCES

Astor, G. (1993). Battling Buzzards. New York: Dell Publishing.

Archer, C. (1985). Paratrooper' Odyssey, A History of the 517th Parachute Combat Team. Florida: 517th Parachute Regimental Combat Team Association.

Barrett, B. (2006). Email message on 517th PRCT email site. "H" Company 517th PRCT.

Biddle, M. (2006). Interview with Mr. Melvin Biddle, March 17, 2006

Breuer, W. (1987). Operation Dragoon. Novato CA: Presidio Press.

Brissey, E. (2006). Email message on 517th PRCT email site. "E" Company 517th PRCT.

Boyle, W. (2003), Email message on 517th PRCT email site

Boyle, W. (2007), Interview, 517th PRCT Association Reunion, June 30, 2007, Washington, D.C.

Clark, B. (1982). Interview about 106th Infantry Division. Fort Knox, KY

Cross, T. (2004). Official After action report of Operation Dragoon, 517th PRCT Association web site: www517prct.org

Del Calzo, N and Collier, P. (2003). Medal of Honor, Portraits of Valor Beyond the Call of Duty. New York, Artisan, Division of Workman Publishing,Inc.

Delesio, G. (2008). Interview, 517th PRCT Association Reunion, June 27, 2008, Saint Louis, MO.

De Trez, M. (1998). First Airborne Task Force, Pictorial History of the Allied Paratroopers in the Invasion of Southern France. Belgium: Publie Par D-Day Publishing.

Flanagan, E. (2002). Airborne, A Combat History of American Airborne Forces. New York: Random House, Ballantine Publishing.

Fraser, D. (2004). Email message on 517th email site "HQ" First Battalion 517th PRCT

Fraser, D. (2006). Interview about gloves of Bud Curtis. "HQ" First Battalion 517th PRCT, Reunion, Portland, Oregon, July 20, 2006

Graves, R. (1985). Letter concerning the men of the 517th, recorded in Paratrooper Odyssey, page v.

Graves, R (1944). Top Secret Letter to men of 517th PRCT, 11 August 1944, Headquarters 517th Parachute Infantry, CT, APO 758, U. S. Army

Hensleigh, H. (2005). Email message on 517th PRCT email site. "HQ" Third Battalion, 517th PRCT

Hensleigh, H. (2009). Email message on 517th PRCT email site. "HQ" S-2, Third Battalion, 517th PRCT

Houston, W. (2005). Article in Thunderbolt Newspaper, April 2005, page 39.

La Chaussee, C. (2005). Article in Thunderbolt Newspaper, first quarter, page 41

McAuliffe, A. (1944). Christmas Letter to the men of the 101st Airborne Division. 24 December 1944.

Robenblum, M. (2004). Invasion of Southern France August 15,1944, Associated Press, May 16, 2004.

Ruggero, E. (2003). Combat Jump. New York: Harper Collins Publishing.

Saunders, D. (2006). Email message on 517th PRCT email site. 596th Engineer Company 517th PRCT.

Schofield, E. (1985). Comments from Paratrooper Odyssey, page 19.

Seitz, R. (2007). Interview, 517th PRCT Association Reunion, July 1, 2007, Washington, D.C.

Smith,J. (1981) Doctrine and Covenants. Corporation of the President of the Church of Jesus Christ of Latter Day Saints. Salt Lake City, UT

Walsh, T. (2005). Email message on 517th PRCT email site.

Webster, D. (2002). Parachute Infantry. New York: Dell Publishing

Wheeler, R. (2008). Interview, 517th PRCT Association Reunion, June 27, 2008, Saint Louis, MO.

INDEX

442nd Regimental Combat Team............124
82nd Airborne Division............190, 196, 199, 212-213, 224, 235
13th Airborne Division.............133, 207, 219, 221-222, 224-225, 229, 233, 235-236
17th Airborne Division...........12, 77, 88, 224, 233-235
36th Infantry Division 123, 132, 224
Adolph the cat 163, 252
Airborne Creed..................................50, 267
Anvil..126, 130
AWOL...... 40, 67-68, 73-74, 99, 101, 162, 170
Bagnoli.. 118-119, 148
Basic training 19, 23, 25-33, 38, 52
Battle of the Bulge .. viii, 75, 88, 92, 177, 186, 194-196, 198, 206, 211-212, 217, 240
Blue Army.......................................77, 88, 91
Bob Douglass...2, 158
Bob Hope............................... 22-23, 125
Bob McMahon...201
Bob Steele 60, 201
Bronze Star 207-208, 226, 229-230, 232
Camp MacKall, North Carolina, ... 67-68, 96
Camp Toccoa, Georgia 12, 15, 19, 22, 32-33, 112, 120
Carl Starkey...200
Charles Diegh 105
Cigarettes..128, 216
Civitavecchia.............................. 123, 148-149
Clickers 153
Col des Braus, France 163
Dean Hildreath..... 39, 41, 44-45, 75, 78, 158, 197, 218, 225-226
Dean Taylor.....................................25, 38, 178
Demolitions Platoon 15
Divisional ABC Test................................63
Don Fraser 72, 112, 194, 196
Drop Zone 143, 149
Eddy Hunter 5
Field problem................. 60, 68, 70, 84, 93-94
First Airborne Task Force, ...224, 263
Fort Benning, Georgia 29, 48-50, 52-56
Garth Boyce....................................4, 114, 178
Herbert Bowlby 142
Howard Hensleigh...... 12, 198-199, 224, 236
Jack Dunaway..97, 220
James Pease 106
Jay Littlefield..196
Jerry Colonia 23
Joe Sumptner..................... 148-149, 200, 206
John Alicki..13
John Lissner.......................................13, 112
John Marsinko 206
JROTC..3
Jump Boots 7, 36-37, 57
Kitchen Police ... 14, 29, 47-48, 61, 69, 79, 81, 97, 101
Krauts.. 170, 173, 195, 209, 217, 219, 230-231
Legion of Honor Medal.........................2, 245
Louis Walsh 12, 236
LST...123, 130
Malmedy Massacre........................... 211-212
Melvin Biddle 188-189, 263
Mock Tower 18
Myrle Traver 213-214
Naples, Italy 118-119, 123, 148
Nice, France...148
Obstacle Course.............................22, 31, 61
Operation Burma...74
Operation Choker 221
Operation Dragoon.. 126, 129-131, 133, 135, 144, 147-148, 263, 267
Operation Overlord 126, 130
Operation Varsity......................................221
Parachute jump........ix, 39, 76, 128, 155, 221, 255
Physical Training...39
Pollywogs..38
Red Army..88, 91

Retreat 105, 145, 189-190, 203
Richard Seitz 14, 142
Rifle Range 32-33
Roger Homer 2
Rupert Graves 112, 143, 236
Sospel, France 251
Stavelot, Belgium 198
Tennessee Maneuvers .... vii, 85, 87-90, 92-93, 123, 131, 187-188, 207, 209, 236
U.S.O. 46, 67, 94, 125
WAACS 37, 54
Willard Hill 7-8, 17, 19, 30, 119, 123, 197, 205
William Boyle 142
Wills 95
XVIII Airborne Corps 177, 186, 190, 212, 219

ABOUT THE AUTHOR

L. Vaughn "Lory" Curtis is the oldest of the three sons of Harland "Bud" Curtis. Raised in California, Lory remembers his paternal grandmother reading his father's letters from the war to him. She explained the terrible things his father experienced as a World War II combat paratrooper. Lory never forgot the stories, and always admired and respected his father's military service. Upon completing high school, Lory joined the United States Marine Corps, serving ten years with service in Vietnam. He then obtained his bachelor's and master's degrees from San Diego State University. Lory taught vocational education and was an associate professor at Palomar College. Lory later returned to active, reserve, and National Guard service in the United States Army as a helicopter pilot. There he served seventeen years. Lory graduated from the Command and General Staff College, and later completed postgraduate studies at the University of Utah in Education Administration, receiving his endorsement as an education administrator. He received his doctorate in Educational Leadership and Foundations from Brigham Young University. Lory has served for nineteen years as a secondary principal.

Upon rediscovering his father's letters in 2003, Lory became driven to compile these letters into a book. Being an avid reader of military history, Lory took each letter, and typed them, giving context using documented history of the 517th Parachute Regimental Combat Team (PRCT), other accounts from 517th paratroopers, and books about World War II paratroopers, establishing the wonderful account, Letters Home: A Paratrooper's Story. From 2004 until 2007, Harland Curtis edited the manuscript and ensured his story was correct.

Letters Home: A Paratrooper's Story is not an account of what Harland Curtis remembered, but are actual letters, describing what happened to him on each specific day. The story entails his adventures as a boy at home and his travels around the western United States, prior to entering military service. It also gives an account of his love for Jill, his girlfriend, and later, fiancée; his rugged and tough paratrooper training of 1943, and his combat experiences from 1944 to 1945.

Letters Home: A Paratrooper's Story, inspired the movie, Saints and Soldiers Airborne Creed (see airbornecreed.com), a movie about 517th paratroopers, by Producer Adam Able and Director Ryan Little. Jasen Wade portrays Harland "Bud" Curtis and depicts the paratroopers' participation in Operation Dragoon, the invasion of South France on August 15, 1944.

Along with articles published in educational journals, Lory is the author of two other books, The Making of an American (2009), Deseret Sun Publishing, and Automotive Diagnosis and Tune-up (1984), McGraw Hill Book Company.

Lory remains committed to preserving the history of the men of the 517th PRCT and other military war veterans. Lory serves as the Vice President of the 517th Parachute Regimental Combat Team (PRCT) Association, and is a professional motivational speaker and can be contacted by email at drlvc69@gmail.com.